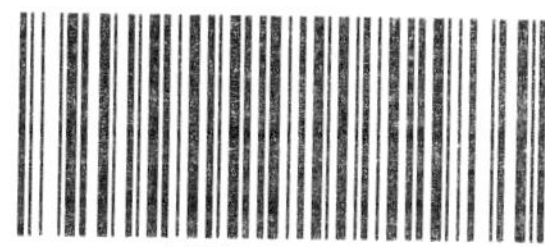

Contents

About This Book

Purpose

This book teaches you how to access Medicare data and, more importantly, how to apply this data to your research. Knowing how to use Medicare data to answer common research and business questions is a critical skill for many SAS users. Due to its complexity, Medicare data requires specific programming knowledge in order to be applied accurately. Programmers need to understand the Medicare program to properly interpret and use its data. With this book, you'll learn the entire process of programming with Medicare data—from obtaining access to data; to measuring cost, utilization, and quality; to overcoming common challenges. Each chapter includes exercises that challenge you to apply learned concepts to real-world programming tasks.

Is This Book for You?

This book is a must for anyone interested in programming with Medicare data in SAS. This includes students, professional researchers, and employees in the federal and state governments. *SAS® Programming with Medicare Administrative Data* offers beginners a programming project template to follow from beginning to end. It also includes more complex questions and discussions that are appropriate for advanced users.

Prerequisites

This book has no prerequisites, and is designed for SAS users at all levels, as well as readers with varying levels of knowledge of the Medicare program and Medicare data. Those readers with experience in SAS, but with no background in healthcare or Medicare data, could focus more on Chapter 1 through Chapter 5. In these chapters, we learn about the Medicare program, how the Medicare program drives the content of the data, and the acquisition of Medicare administrative claims and enrollment data. Those readers with knowledge of the Medicare program, but not Medicare data or SAS, could focus more on Chapter 3, where we discuss Medicare data, as well as Chapter 6 through Chapter 10, where we develop the majority of SAS code.

Scope of This Book

The main topics covered in the book include:

- An introduction to the Medicare program;
- A detailed introduction to Medicare data;
- A discussion of how to plan a research programming project;
- Methods to obtain Medicare data;
- Instructions for programming with Medicare data to measure utilization, cost, and quality;
- A demonstration of how to present findings;
- A helpful explanation of how to sunset a research programming project

Because this book is foundational, it does not cover more complex manipulations or analyses, such as risk adjustment or payment standardization algorithms. It does, however, provide the reader with ample background information to tackle these topics independently.

About the Examples

Software Used to Develop the Book's Content

SAS 9.3

Example Code and Data

You can access the example code and data for this book by linking to its author page at http://support.sas.com/publishing/authors/gillingham.html. For an alphabetical listing of all books for which example code and data is available, see http://support.sas.com/bookcode. Select a title to display the book's example code.

If you are unable to access the code through the website, e-mail saspress@sas.com.

Output and Graphics Used in This Book

The output in this book was generated using ODS.

Exercise Solutions

You can access the exercise solutions for this book by linking to its author page at http://support.sas.com/publishing/authors/gillingham.html.

Keep in Touch

We look forward to hearing from you. We invite questions, comments, and concerns. If you want to contact us about a specific book, please include the book title in your correspondence to saspress@sas.com.

To Contact the Author through SAS Press

By e-mail: saspress@sas.com

Via the Web: http://support.sas.com/author_feedback

SAS Books

For a complete list of books available through SAS, visit http://support.sas.com/bookstore.

Phone: 1-800-727-3228

Fax: 1-919-677-8166

E-mail: sasbook@sas.com

SAS Book Report

Receive up-to-date information about all new SAS publications via e-mail by subscribing to the SAS Book Report monthly eNewsletter. Visit http://support.sas.com/sbr.

Publish with SAS

SAS is recruiting authors! Are you interested in writing a book? Visit http://support.sas.com/saspress for more information.

About The Author

Matthew Gillingham is the director of health research systems at Mathematica Policy Research. His career in health care data analytics began with analyzing commercial claims and enrollment data. For more than ten years, he has used SAS to build software that analyzes administrative health care and Medicare administrative data. Specifically, Gillingham directs systems research and development work that uses large Medicare claims and enrollment data files to monitor and report on health care quality and utilization for the Centers for Medicare & Medicaid Services. In addition, he provides technical assistance and quality assurance on projects that use Medicare data to create analytic files for program evaluation, monitoring, and other research purposes.

Learn more about this author by visiting his author page at http://support.sas.com/publishing/authors/gillingham.html. There, you can download free chapters, access example code and data, read the latest reviews, get updates, and more.

Acknowledgments

Writing this book became more of a lifestyle choice than a project (kind of like owning a boat, but without the enjoyment of being on a boat!). What's more, this "lifestyle choice" was not just my own. Family, friends, and coworkers provided support and patient encouragement as they endured my long hours, late nights, and busy weekends with grace.

I would like to thank my employer, Mathematica Policy Research, as well as my colleagues. I consider myself truly blessed to work at Mathematica, a company that values the highest standards of quality, unwavering objectivity, and compassionate collegiality. My years at Mathematica have been filled with opportunities to perform interesting and relevant work, interact with some of the brightest minds in policy research, and benefit from supportive, caring, and compassionate leadership. I simply cannot imagine working anywhere else.

I would like to thank my reviewers at Mathematica: Sandi Nelson, Jeffrey Holt, Scott McCracken, and Ken Peckham. Their thoughtful reviews and commentaries, as well as their support and friendship, made this book much better than its original conception. In addition, I owe a debt of gratitude to Scott Greiner, Patrick Wang, Ken Peckham, and Scott McCracken, who assisted me with the creation of dummy data, and Susie Clausen, who did a wonderful job formatting my written materials. Carmen Ferro and Mathematica's communications department acted as a creative force in the promotion of the book and the book's cover design. I would like to thank John West, my editor at SAS Publishing, for his patience and guidance. Importantly, I would also like to thank the book's technical reviewers, Gerri Barosso, Russ Tyndall, John Shipway, Bernadette Johnson, Craig Dickstein, and Deidra Peacock, for their thorough and thoughtful review of my manuscript.

My supervisor, Edward Hoke, provided supportive guidance and encouragement. Ed certainly cannot fathom what a wonderful influence he has been in my life. Senior leaders at Mathematica, Dr. Myles Maxfield, Dr. Craig Thornton, Adam Coyne, Pam Tapscott, Tamara Barnes, Andrea Darling, Patricia Guroff, and Mary Harrington, not only provided encouragement, but helpful guidance as well. Finally, my staff at Mathematica's Ann Arbor office (Dean Miller, Ken Peckham, Kelly Zidar, Kevin Bradway, Gregory Bee, Rachel Thompson, Gerry Skurski, Anne Harlow, and Nate Darter), as well as close colleagues like Justin Oh, offered encouragement, support, and guidance.

Friends and family supported me and helped me see the project through to completion. My mom, dad, and sister taught me to believe in myself and dream big dreams. Anne, Owen, and Lucy provided a safe, loving, and nurturing place to rest and relax after long workdays. Anne, in particular, never let me forget that I could see this project through to the end. Lastly, my good friend Chris Thieme not only provided encouragement, but he also taught me much of what I know about SAS (although any mistakes are my own!). He will always be the most skilled, and coolest, SAS programmer I know.

These people allowed me to achieve a lifelong dream. I owe them a debt that I will never be able to repay. I hope they take some pride of ownership of this project, because it is just as much their project as it is mine.

Chapter 1: Introduction

Introduction and Purpose of This Book

Welcome to the world of SAS® programming with Medicare data! This book provides beginner, intermediate, and advanced SAS users with the information needed to execute a research programming project using Medicare administrative data. It introduces the reader to common, important, and frequently used concepts encountered when programming with Medicare data. The focus of this book is decidedly on the data and the policies that drive the content of the data, using the SAS syntax as a tool to answer common research questions. This approach is very intentional; in my experience, the SAS syntax is the easiest part of programming with Medicare data! The most challenging aspect of my job is to understand the data and how to properly use them to correctly answer research questions.[1]

Medicare data are unlike any other health care data, and the uniqueness of Medicare data is driven by the uniqueness of the Medicare program. For example, public policy decisions largely influence decisions surrounding what services Medicare covers and how those services are paid for. This differs from the decision making process of a commercial health plan, which seeks to maximize profits. Here are a few examples, some of which we will discuss in coming chapters:

- Medicare generally covers elderly beneficiaries without regard to medical history. Therefore, the Medicare population is generally sicker than the population of most commercial health plans.
- Part D prescription drug data contains information for out-of-hospital prescription drug fills. Prescription drugs administered during a hospital stay may not appear in the claims data at all.

- Services paid for by Medicare Part C (Medicare Advantage) may not appear in the administrative claims files because they are paid for by managed care providers.
- Medicare pays for some services (e.g., home health agencies, hospice, or acute inpatient hospitals) using what are called prospective payment systems (PPS).

These examples are supposed to whet your appetite for learning about Medicare data. I am hoping that you have not run screaming from your desk! If you have, please come back! We will learn about all of these concepts and more! The point I am trying to make is that the more you understand the Medicare program, the more you will understand how to properly use Medicare administrative data for research purposes. To this end, this book will describe the Medicare program, Medicare data, how the two interact, and how that affects the SAS programmer who uses the data. These concepts and techniques will be illustrated through the completion of an example research project.

Learning how to use Medicare data is a lifetime pursuit. It takes many, many years to build subject matter expertise due to the sheer volume of information on topics like Medicare policy, payment systems, and data. What's more, the world of Medicare data is changing rapidly due to increased attention on Medicare costs, quality, and the use of information technology in health care delivery. For example, the use of electronic health records (EHRs) in the measurement of utilization, cost, and quality is on the horizon. EHRs will supplement, or perhaps someday replace, the administrative claims data we use in this book. In addition, Medicare is exploring different methods of payment, such as the bundling of services previously paid on an individual basis. As such, we will never be able to cover every topic you will run into. However, we will build a very strong foundation in standard concepts, such as specifying and coding continuous enrollment algorithms and identifying and summarizing common services and events. My hope is that the reader can apply this foundation to programming projects that use Medicare data and use it to gain a broader perspective on the use of health care administrative data in general.

Framework of This Book

This book is organized as an approach to completing a research programming project. The basic framework below can be used as a blueprint for programming most any research project, including the example we will use in this book.

1. Plan the project by identifying the research questions to be answered and thinking through how we will construct our code to provide those answers. Prior to writing any code, we will create a data flow diagram of our programming plan and use that plan to request data. Typically, prior to writing any SAS code you would write programming specifications (or business requirements or functional requirements) that state exactly how you plan to answer the research questions using SAS code. For illustrative purposes, we will write the specifications as we code (i.e., the specifications will be our explanation of the code).

2. Obtain the needed data. There are many different types of data you can use, depending on the nature of the project and its requirements. Our project will focus on using administrative claims and enrollment data.
3. Develop code to create the analytic files needed to answer research questions. Analytic files are essentially summaries or subsets of the raw source data that we will use to perform the analysis that answers the research questions.
4. Develop code that utilizes the analytic files to create answers to the research questions.
5. Perform quality assurance and quality control of our algorithms.
6. Run our algorithms in production, typically by using a batch submittal.
7. Create documentation, take steps to preserve our output data sets, and complete any contractually required data destruction.

Our Programming Project

During the planning process for this book, I thought long and hard about an example project that would be useful to a variety of users. I wanted the project to address common research questions and result in the creation of algorithms that are almost universally applied to research programming work that uses Medicare data. As such, I came up with the following criteria for the project:

- The research questions must be applicable and relevant to today's research environment. For example, the accurate measurement of utilization and cost of health care services are topics that have been consistently important and relevant since the first person used a computer to analyze health care claims.
- The rcsearch questions must lend themselves to building a foundation for addressing "real world" questions. The foundation for all research programming projects is the study population, so our example project will include obtaining beneficiary enrollment data for defining continuous enrollment.
- The research questions must result in algorithms that are easily adaptable to being used to answer research questions in your "real world" work. For example, we will create an algorithm that defines continuously enrolled beneficiaries as those beneficiaries who have had Medicare Fee-for-Service (FFS) coverage for all 12 months of a study year. This kind of algorithm is easily modified to define continuously enrolled beneficiaries as those beneficiaries enrolled in Medicare Advantage (MA) for 6 months of the year.
- Although this is an introductory text, the research questions must illustrate some of the complexity of using Medicare data. If there is one common trait that unites all of the projects that I have worked on, it is that the project always grows in complexity.

With these criteria in mind, I designed the following example research project:

Let's imagine that we are working at a university or a policy research company (maybe you don't need to imagine this because you already are!). Let's further imagine the Centers for Medicare & Medicaid Services (CMS) enacted a pilot program during calendar year 2010[2] designed to reduce costs to Medicare and improve (or at least not harm) quality outcomes for Medicare beneficiaries. The details of the pilot program are not particularly important to our effort, so let's imagine that the program provides an extra payment to providers that significantly reduce payments and improve quality outcomes when compared to groups of their peers. We have been asked to evaluate the effectiveness of the program, and we would like to start by measuring simple payment, utilization, and quality outcomes for those providers that interacted with the beneficiaries in our sample population during the study year 2010. Because the program had been operational for the full 2010 calendar year, we can identify the providers that participated in the pilot. Therefore, the starting point for our example research programming project is a file provided by CMS that contains identifiers for the providers that participated in the pilot program, along with identifiers for the beneficiaries associated with those providers.[3] Therefore, we must acquire enrollment and claims data for these beneficiaries, and subsequently develop algorithms that will query the data to produce summaries of payment, utilization, and quality outcomes during the study year. These summaries will be used in our evaluation of the program.[4]

In the end, the goal of this text is not to make any real determinations about utilization, payments, and quality of care; after all, we are using fake data that cannot be used for drawing any real conclusions and exists solely to develop code.[5] Rather, our goal is to prepare you for working on your own real world research projects by using our example research programming project to teach the mechanics of using Medicare data to measure utilization and Medicare payments, and to identify chronic conditions and commonly used indicators of quality outcomes. Therefore, by the end of this text, the reader will understand important concepts that are applicable and foundational to using Medicare administrative claims and enrollment data to, say, identify most any chronic condition or compute most any quality outcome metric.

We can now be more specific about the things we would like to measure. In particular, evaluating the success of the program involves coding the following measurements of utilization and payment for the beneficiaries in the pilot program we are studying. We need to:

- Calculate the number of evaluation and management (E&M) visits in a physician office setting, and the amount paid for those E&M services.
- Calculate measures of inpatient hospital utilization, and the amount paid for inpatient hospital claims.
- Calculate the utilization as it pertains to the professional component of emergency department (ED) visits.

- Calculate the utilization of ambulance services.
- Calculate the number of outpatient visits, as well as skilled nursing facilities (SNFs), home health agencies (HHAs), and hospice care.
- Calculate the total Medicare amount paid for all Part A claims for our population.

In addition, evaluating the success of the program also entails coding the following measurements of quality outcomes, often at the physician level:

- Measure evaluation and management utilization for beneficiaries with diabetes or chronic obstructive pulmonary disease (COPD).
- Identify the extent to which diabetics received services for eye exams.
- Calculate the number of hospital readmissions for beneficiaries with COPD.
- Finally, we will provide examples of methods to summarize and present results by beneficiary demographic characteristics, as well as by provider. While these examples are by no means exhaustive (e.g., we do not summarize and present every analysis performed in earlier chapters, we do not endeavor to analyze results using a control population, and we do not look for significant changes in performance over time), they do provide the reader with a foundation for further work.

The above concepts meet our criteria of being relevant, foundational, and adaptable. For example, instead of studying hospital admissions for Medicare beneficiaries with diabetes, you could study the same utilization and cost measurements for beneficiaries with prostate cancer. Similarly, you could adapt the measurement of retinal eye exams for diabetics to examine a different procedure (say, immunization for influenza) for beneficiaries with a different chronic condition (say, beneficiaries with prostate cancer or COPD).

Chapter Outline

Each chapter in this book will address a section of the project:

- Chapter 2 sets a foundation for using and understanding the data by learning about the Medicare program. Remember, the guiding principle of this book is that the only proper way to answer research questions about the Medicare program is by understanding the program that drives the data.
- Chapter 3 builds on the foundation developed in Chapter 2 by describing the content of Medicare data files in detail.
- Chapter 4 plans the project by describing the initiation, planning, and design phase of the Systems Development Life Cycle (SDLC).

- Chapter 5 covers requesting, obtaining, and loading the necessary data. In this chapter, we begin to work with our source data. This chapter marks the beginning of the discussion on the creation of the analytic files we will use to summarize our results and ultimately to answer our research questions.
- Chapter 6 defines beneficiary enrollment characteristics, including the creation of variables that indicate continuous enrollment, age, and geographic information.
- Chapter 7 presents code to calculate the aforementioned measurements of utilization.
- Chapter 8 presents code to calculate the aforementioned measurements of Medicare payment.
- Chapter 9 identifies common medical conditions by focusing on diabetes and COPD, and develops examples of basic measurements of quality outcomes for beneficiaries who have these conditions.
- Chapter 10 focuses on bringing the output of Chapter 5 through Chapter 9 together, using that output for answering the research questions, and presenting those answers. We will also discuss the steps involved in finalizing our work through documentation, preservation of code, and complying with all of CMS's Data Use Agreement[6] policies, such as the destruction of data.

As you can see, another way of presenting the organization of this book is to say that the first four chapters are not focused on writing SAS code. Rather, they are focused on learning about the Medicare program, Medicare data, CMS's systems, and the unique process of planning a research programming project that utilizes Medicare administrative data. Again, this is intentional. It is significantly important that the reader acknowledge that one must understand the Medicare program in order to successfully work with Medicare data, and one must understand Medicare data in order to properly answer research questions with Medicare data. Therefore, the first four chapters set up a foundation for the remaining chapters, with the coding and the actual execution of answering our example research questions occurring in Chapter 5 through Chapter 10.

How to Use This Book

With this in mind, each reader will come to this book with different levels of programming experience and various levels of exposure to working with Medicare administrative data. I recommend the following approach based on the reader's experience:

- Those readers with experience in SAS, but with no background in healthcare or Medicare data could focus more on Chapter 1 through Chapter 5, where we learn about the Medicare program, how the Medicare program drives the content of the data, and the acquisition of Medicare administrative claims and enrollment data.
- Those readers with knowledge of the Medicare program, but not Medicare data or SAS could focus more on Chapter 3, where we discuss Medicare data, as well as Chapter 6 through Chapter 10, where we develop the majority of code.

For those readers using the code developed in this book, it is important to note that there will almost certainly be ways of making the code we write more efficient. I have consciously sacrificed developing code that processes efficiently (e.g., shorter CPU and wall time) for the efficiency gained by writing code that clearly steps through a process, even if it involves coding additional DATA steps. Indeed, the full set of algorithms presented in Chapter 5 through Chapter 10 can be combined into fewer steps requiring less reading and writing of large Medicare claims files, which would result in a set of code that processes faster. However, our objective is to learn about programming with Medicare data, so we are sacrificing those efficiency gains in order to develop specific algorithms in a stepwise fashion, chapter by chapter. It would be a good exercise for you to revamp the code developed in Chapter 5 through Chapter 10 in order to reduce processing time (and I'd love to see your results!).

The online companion to this book is at http://support.sas.com/publishing/authors/gillingham.html. Here, you will find information on creating dummy source data, the code presented in this book, and answers to the exercises in each chapter. I expect you to visit the book's website, create your own dummy source data, and run the code yourself.

Disclaimer

The synthetic data used for purposes of this book originated with the Centers for Medicare & Medicaid Services' (CMS) Data Entrepreneurs' Synthetic Public Use File (DE-SynPUF), which is available in the public domain. While the DE-SynPUF is derived from data that is used by CMS for operational purposes, the DE-SynPUF does not permit direct identification of any individuals because all direct identifiers have been removed. The author assumes no responsibility for the accuracy, completeness, or reliability of the DE-SynPUF, and assumes no responsibility for the consequences of any use of the data or algorithms contained in this book. The data are used herein without any representation or endorsement and without warranty of any kind, whether express or implied. IN NO EVENT SHALL THE AUTHOR BE LIABLE FOR ANY DIRECT, INDIRECT, SPECIAL OR INCIDENTAL DAMAGES RESULTING FROM, ARISING OUT OF OR IN CONNECTION WITH, THE USE OF THE DATA OR ALGORITHMS CONTAINED HEREIN.

Chapter Summary

In this chapter, we introduced the purpose of the book, described the framework of the book, and specified our example research programming project.

---

[1] The idea of understanding Medicare data prior to writing any code is so important that the first five chapters of this book are focused on laying out the framework of the book, learning about the Medicare program, Medicare data, and CMS's systems, and the unique planning process of a research programming project that utilizes Medicare administrative data. We do not begin to write any SAS code until Chapter 5!

[2] Updating the year of study requires examining the choice of descriptive codes (such as procedure codes) discussed in later chapters. The need to choose relevant codes based on year of study is very common in health services research. In Chapter 10, we present an exercise that asks the reader to contemplate the changes we would need to make to update the text as if the demonstration program we are researching took place during the year 2015.

[3] We will discuss the use of this file more in Chapter 5. Specifically, this file will serve two purposes in our work. First, we will use this file as a "finder file" that serves as the basis for our data extraction. In addition, we will use this file to assign responsibility for a beneficiary's care to a provider (called attribution). We assume that this finder file was sent to us as a SAS data set and a flat text file.

[4] It is worth noting that these types of evaluations are not unique to the Medicare world. Evaluative studies like the sample project we will undertake in this book are frequently performed by private health plans, government purchasing agents, and other entities around the world.

[5] For more information, please see the disclaimer below.

[6] A Data Use Agreement (DUA) is a contract governing how the user will interact with the data, including data security and data destruction procedures.

Chapter 2: An Introduction to the Medicare Program

Introduction and Goals of the Chapter

This chapter presents an introduction to the Medicare program and is designed to prepare the SAS programmer for using Medicare administrative data. As stated in Chapter 1, the guiding principle of this book is that research questions about the Medicare program can only be answered with a solid understanding of the fundamentals of Medicare data. In turn, Medicare data can only be understood when the user achieves a solid grasp of the fundamentals of the Medicare program. Indeed, as is true with most types of administrative data, it is the requirements of the program that drive the content of the files. In that spirit, the goal of this chapter is to establish a foundation for understanding and using Medicare data by learning the basics of the administration of the Medicare program. We define Medicare, discuss enrollment, eligibility, and coverage, and provide a very simple sketch of how Medicare pays for services. We also briefly discuss how this information about Medicare influences the content of the data files we will use throughout the remainder of this book. Looking forward, we will build on the information presented in this chapter by discussing more specifics of Medicare data files, as well as how to request, obtain, and use these files. We will use the data in these files to address the research questions posed by our example project described in Chapter 1.

It is very important to note that Medicare coverage is extremely complex and subject to change over time. We do not attempt to cover every detail of the Medicare program in this text. You will encounter many questions throughout the course of your career using Medicare data that will require you to dig deeply through reference material. To that end, the intent of this chapter is to provide a foundation for understanding the Medicare program for your future work. My hope is that the reader finishes this chapter with a basic understanding of the Medicare program, including history, types of coverage, and administration. When confronted with more advanced research questions, the reader can then leverage this understanding in conjunction with available reference materials. Some of the most commonly used sources of information are CMS's Research Data Assistance Center (ResDAC), CMS, and the Kaiser Family Foundation.[1] In fact, these sources were heavily relied upon in the writing of this chapter!

An Introduction to the Medicare Program

What is Medicare?

Medicare is a health insurance program for people age 65 or older, those under age 65 with certain disabilities, and those of any age with permanent kidney failure. The Medicare program protects beneficiaries from financial risk by covering costs for potentially large and unaffordable medical expenses incurred by seeking medical care. Generally, in order to be eligible for Medicare, beneficiaries must have entered the United States legally, paid Federal Insurance Contributions Act (FICA) taxes for 40 or more quarters (or be the spouse of someone who has), and lived in the United States for 5 years.[2] Medicare is a social insurance program operated by the Centers for Medicare & Medicaid Services (CMS), a federal government agency that is part of the Department of Health and Human Services. Medicare provides participants (called beneficiaries) with an array of health insurance coverage, regardless of income or medical history. Medicare provides four types of coverage (Part A, Part B, Part C, and Part D) that are described in detail below.

Started in 1965 (did you know that President Harry Truman was the first person to enroll in Medicare?), the Medicare program we know today (and will describe below), had its genesis in President Lyndon Johnson's War on Poverty. Because Medicare is a social insurance program, enrollment criteria and benefits are defined by legal statute. This means that Medicare coverage can differ from commercial health insurance in some fundamental ways. It also means that Medicare has changed over the years in response to changes in statute, often to expand or improve coverage or to attempt to control costs. Here are just some examples of how legislation has influenced the administration of the Medicare program:[3]

- In 1972, the Medicare program was expanded to include coverage for individuals with end-stage renal disease (ESRD) and some individuals under age 65 with long-term disabilities.
- In the same year, coverage was also expanded to include speech, chiropractic, and physical therapy services.
- In 1982, Medicare coverage was expanded to include hospice services for terminally ill individuals.
- In 1997, the Balanced Budget Act attempted to control Medicare spending through the creation of prospective payment systems (PPS) for certain types of services (though inpatient prospective payment was first implemented in 1983), and established the Medicare+Choice program.
- In 2001, Medicare initiated coverage for individuals with Lou Gehrig's disease (ALS).
- In 2003, The Medicare Prescription Drug, Improvement, and Modernization Act (MMA) established an outpatient prescription drug benefit that would take effect in 2006.
- In 2005, coverage was expanded to include a physical and preventive screening to new Medicare beneficiaries.

- In 2010, the Affordable Care Act (more commonly known as "health reform legislation") initiated sweeping measures to control costs, most of which will take effect by 2014. For example, the law provides increased funding to combat waste, fraud, and abuse, takes measures to attempt to improve the quality of care provided to beneficiaries, and establishes free annual wellness visits for Medicare beneficiaries.

Medicare Enrollment and Eligibility

At the time of this writing, Medicare provided health insurance to about 47 million Americans.[4] Most people think of Medicare as insuring the elderly, and that is certainly true; the majority of Medicare beneficiaries (about 39 million of them) are eligible for Medicare insurance because they are aged 65 and over. However, Medicare also insures about 8 million beneficiaries who are permanently disabled (receiving Social Security Disability Insurance, or SSDI), have end stage renal disease (ESRD, a condition that requires dialysis), or ALS, regardless of their age. You may hear experts refer to beneficiaries aged 65 and older in general terms as "aged," and those under age 65 as "disabled."[5]

What Is Covered by Medicare?

Medicare benefits are divided and defined in four parts (Part A, Part B, Part C, and Part D). Each Part covers a different type of care or set of services. As we will see in subsequent chapters, not only are these Parts a way of describing coverage, but also a way of organizing the administrative data files we will use throughout this book. Understanding Medicare coverage (and limitations to that coverage) is essential to the proper utilization of Medicare claims data. For example, let's say you were asked to study claims for blood received in a transfusion. Medicare Part A covers the blood received by a beneficiary in an inpatient hospital setting, but Medicare Part B covers the blood the same beneficiary may have received in a hospital outpatient setting. This means that the programmer may need to query more than one dataset to locate blood-related information in the claims data. As we will see, querying more than one type of claims data set is important in the identification of emergency department visits.

The specifics of Medicare coverage are subject to, and often do, change. As such, it is very helpful to be able to tap into reference materials that summarize Medicare benefits. As mentioned above, Medicare is a social insurance program and the final source of information on Medicare coverage is legal statute. However, many experts simply refer to summaries of the Medicare schedule of benefits that CMS provides to beneficiaries, including online publications such as Your Medicare Benefits[6] and Medicare and You[7]. These publications were used as the foundation for some of the information presented below.

- Medicare Part A, also known as Hospital Insurance (HI), pays for care provided to beneficiaries in hospitals (including most inpatient care, inpatient rehabilitation facilities, and long-term care hospitals), coverage for short-term stays in skilled nursing facilities (SNFs), most post-acute care provided in home health agencies (HHAs), and hospice care services.
- Medicare Part B is also known as Supplemental Medical Insurance (SMI) because it provides coverage that is additional and supplemental to Medicare Part A coverage. Part B covers all medically necessary professional services, be they in an inpatient, outpatient, or physician office setting, including visits to the physician, outpatient care, outpatient mental health care, diagnostic and clinical laboratory testing, and some preventative services, like flu and pneumonia

vaccinations. In addition, Part B coverage includes durable medical equipment (DME). The vast majority of beneficiaries with Part A coverage also purchase Part B coverage. Taken together, Medicare Parts A and B are also known as "original fee-for-service (FFS)," "original Medicare," or "traditional Medicare" coverage.

- Medicare Part C, also known as Medicare Advantage (MA) or managed care, provides Medicare beneficiaries with the option of enrolling in a private insurance plan as opposed to participating in traditional Medicare fee-for-service coverage. Private plans include health maintenance organizations (HMOs), preferred provider organizations (PPOs), private FFS plans, Special Needs Plans, and Medicare Medical Savings Account Plans. These MA plans receive payments from Medicare (and premium payments from members) to provide benefits provided by Medicare Part A (excluding hospice), Part B, and usually Part D. MA plans are required to use extra payments to provide additional benefits, like vision coverage. The number of MA enrollees and plan options has consistently increased since 2004. Beneficiaries have to be enrolled in Part A and B in order to join an MA plan. As noted in Chapter 1, MA claims may not appear in the administrative claims files provided by CMS because they are paid by private managed care insurance plans. Therefore, it is not uncommon for investigators to exclude MA beneficiaries from evaluations similar to our example research programming project.
- Medicare Part D is voluntary prescription drug coverage. In other words, a familiar way to think about Part D is that it helps pay for prescription drugs prescribed by doctors and filled at a pharmacy.[8] The program is relatively new (it was launched in 2006) and helps pay for drugs through private plans, called standalone prescription drug plans (PDPs) and MA prescription drug plans (MA-PDPs).

What Is Not Covered by Medicare?

Like other health insurance plans, Medicare does not cover every possible medical service or procedure. In addition, Medicare may require beneficiaries to make certain cost-sharing payments, like deductibles and coinsurance. Finally, although Medicare may cover the service, Medicare may not be the primary payer for services provided to beneficiaries who carry additional health insurance coverage. Below are some examples of services with limited or no coverage under Medicare. As you will see, a proper understanding of coverage (and limitations) is vital to the accurate identification of services in the administrative data.

- Some services have limitations on coverage. For example, Medicare Part A stops paying for inpatient psychiatric care in a psychiatric hospital after 190 days (this is a lifetime limit)[9], and a beneficiary can only be admitted to a skilled nursing facility after being discharged within the last 30 days from an inpatient hospital stay that lasted at least three days.[10]
- Other services are simply not covered. For example, Medicare does not cover long-term care services (care received in a nursing home, respite care, and adult day care) at all. Also, it does not cover cosmetic surgery, some preventative services (although this is changing with the Affordable Care Act), vision and dental care, and hearing aids.
- Medicare is a secondary payer for beneficiaries that have certain additional health insurance coverage. For example, if a beneficiary has been diagnosed with black lung disease and the beneficiary is covered under the Federal Black Lung Program and Medicare, the Federal Black Lung Program will pay for services related to the beneficiary's black lung condition. In this

case, Medicare is a secondary payer, meaning that it may cover the remainder of the claim not paid by the Federal Black Lung Program but is not responsible for the primary payment of the claim.

Some limited and uncovered services, as well as cost-sharing payments, can be covered by supplemental insurance. Specifically, beneficiaries can acquire supplemental coverage from several sources: Medigap insurance policies, insurance sponsored by their employers, MA plans, and, in some cases, Medicaid. Note that it is very possible that claims are not filed with Medicare for medical services paid for by the beneficiary out-of-pocket or by coverage other than Medicare. As we will see below, this means that the user of Medicare administrative data may not be able to account for all services a Medicare beneficiary receives.

The Mechanics of Medicare

Now that we better understand some of the basics of Medicare and Medicare coverage, we can discuss how covered beneficiaries receive services and how Medicare reimburses providers of those services.

You probably have or have had commercial health insurance of your own, and in some very basic ways it does not operate much differently than Medicare. When you go to the doctor for an examination, the physician that examines you submits a bill (called a claim) to your insurance company for reimbursement. This claim is usually submitted electronically and describes the services provided by the physician (in this case, let's say a routine visit to the doctor for a checkup, called an evaluation and management examination) and the charges for those services. More specifically, the claim describes you (e.g., name, personal identifier, age, and sex), the provider of the service (e.g., name, provider identifier, and place of service), the date or dates of service, and details that describe the services performed, like procedure and diagnosis codes. When the insurance company receives the claim, it goes through an adjudication process whereby payment is determined. After you meet your requirements as a beneficiary by paying your deductible and coinsurance, your insurance company typically pays the remainder of the claim for eligible services (perhaps an amount adjusted to account for negotiated purchasing agreements) according to the terms of your coverage. Your health insurance company is able to pay these bills because it maintains a fund of money reserved for just such purposes. This fund is derived in part from the premium payments made by you and other beneficiaries (and, in the case of for-profit insurance companies, the accumulation of profit). These premium payments are determined statistically by actuaries and take into account the projected risk associated with the level of health of pools of covered beneficiaries.

Medicare operates in many of the same ways as commercial health insurance. When a Medicare beneficiary goes to the doctor for a checkup, the provider submits a bill to Medicare Part B (similarly, if a Medicare beneficiary goes to the hospital, the provider submits a bill to Medicare Part A). Most of this billing is done electronically. The claim form[11] contains details like the beneficiary and provider identifiers, dates of service, place of service, the procedures performed, and the patient's diagnosis. The claim submission will go to the provider's regional Medicare Administrative Contractor (MAC) for adjudication, processing, and payment. Once claims are paid, they are considered final action claims. Final action Medicare claims are stored in files available to the research community as Medicare administrative data. As you may have already inferred, the administrative files that are created from final action claims and enrollment information are derived from systems that are used to administer the Medicare program. In other words, the primary purpose of these systems is not to create data for research, but to adjudicate and pay claims. This fact has implications for using the administrative data files and means that the user must understand the Medicare program. We will explore this topic in detail throughout the remainder of this book. For now, let's end with some well-known examples of how particulars of the administration of the Medicare program influence the content of administrative data files.

- With about 47 million Medicare beneficiaries, we can expect the administrative data files we use to be quite large. As such, we will need to consider efficient programming techniques. Many of the exercises in later chapters address efficiency topics.
- Some services that you want to study may not appear in Medicare administrative data or, at the least, may require searching multiple files. For example, Part D prescription drug data contains information for prescription drug fills. Prescription drugs administered during a hospital stay may not appear in the claims data at all. Additionally, services paid for by Medicare Part C may not appear in the administrative claims files because they are paid for by managed care providers.[12]
- Medicare pays for some services (e.g., home health agencies, hospice, hospital outpatient, skilled nursing, or acute inpatient hospitals) using what are called prospective payment systems (PPS). In very simple terms, a PPS reimburses providers using a fixed amount derived from a predetermined classification system.[13] We will discuss payment systems more in Chapter 8.
- As a social insurance program, Medicare coverage is provided regardless of medical history. Therefore, if you are used to working with commercial healthcare claims data, you will likely notice some unique characteristics of the Medicare population, such as a higher prevalence of chronic conditions.
- The administrative data we use for research purposes are updated on a regular basis, but only with claims that have been received and adjudicated and deemed final action. As such, the files we use at any given time do not contain all final action claims submitted and paid up to the date of extraction of the data. It is common practice to wait at least three months for paid claims to appear in the claims files maintained by CMS. For example, a request for claims for the full calendar year 2014 is best made on or after April 1, 2015.
- Depending on what you are studying, care must be taken to determine the correct composition of your study population. For instance, our example research project will study only those beneficiaries continuously enrolled in fee-for-service Medicare during a defined timeframe. Other studies may wish to focus on beneficiaries entitled to Medicare based on being disabled. We will see in Chapter 6 that we can use enrollment data to determine a beneficiary's reason for entitlement and define our study population.

Chapter Summary

In this chapter, we set a foundation for programming with SAS and Medicare administrative data by examining the following:

- Understanding the Medicare program and the particulars of Medicare coverage is absolutely essential to successfully programming with Medicare administrative data.
- Medicare is a social insurance program that provides beneficiaries with an array of health insurance coverage, regardless of income or medical history.
- The majority of Medicare beneficiaries are eligible for Medicare insurance because they are aged 65 and over. However, Medicare also insures beneficiaries who are permanently disabled (receiving Social Security Disability Insurance or SSDI), have ESRD, or have ALS.
- Medicare benefits are divided and defined in four parts: Part A (Hospital Insurance), Part B (Supplemental Medical Insurance), Part C (Medicare Advantage), and Part D (outpatient prescription drug coverage). Each Part covers a different type of care or set of services. These Parts are a way of describing coverage, but also a way of organizing the way we think about the administrative data files we will use throughout this book.
- Like other health insurance plans, Medicare does not cover every possible medical service or procedure.
- The primary purpose of Medicare payment systems is not to create data for research, but to adjudicate and pay claims. This fact has implications for using the administrative data files and means that the user must understand the Medicare program.

---

[1] See: www.resdac.org, www.cms.gov, www.medicare.gov, and www.kff.org.

[2] Medicare eligibility is more complicated than the simple presentation above. For more information, see federal resources such as http://www.medicare.gov/publications/pubs/pdf/11306.pdf.

[3] See the Medicare timeline at http://kff.org/medicare/video/the-story-of-medicare-a-timeline/.

[4] Information provided in this paragraph, and more, can be found throughout the KFF Medicare Primer (April 2010) (http://www.kff.org/medicare/7615.cfm).

[5] Although we will not discuss Medicaid in this text, some beneficiaries are eligible for and enrolled in both Medicare and Medicaid. These beneficiaries are referred to as Medicare-Medicaid Enrollees, or MMEs.

[6] Your Medicare Benefits can be found at http://www.medicare.gov/publications/pubs/pdf/10116.pdf.

[7] Medicare and You can be found at http://www.medicare.gov/publications/pubs/pdf/10050.pdf.

[8] We mention below that medication provided in institutional settings (like during a hospital stay) may be covered by Medicare, but does not necessarily appear in the administrative claims data files used for research purposes.

[9] The reader should check the Medicare benefits schedule for the most up-to-date information.

---

[10] See CMS's Medicare Coverage of Skilled Nursing Facility Care booklet, page 17, available at http://www.medicare.gov/Pubs/pdf/10153.pdf.

[11] Medicare claim forms include the CMS-1500 for physician or professional billing and the UB-04 (also known as the CMS-1450) for institutional or technical billing.

[12] For more information, see ResDAC's article at http://www.resdac.org/resconnect/articles/114.

[13] Source: CMS's Prospective Payment Systems-General Information website available at https://www.cms.gov/ProspMedicareFeeSvcPmtGen/.

Chapter 3: An Introduction to Medicare Data

Introduction and Goals of This Chapter

This chapter introduces the contents of the Medicare claims and enrollment files we will use for our research programming project. The purpose of this introduction is to build on the explanation of the Medicare program presented in Chapter 2 and further prepare the programmer for using Medicare administrative data to complete our example research programming project. In Chapter 2, we learned the basics of the administration of the Medicare program. We defined Medicare, discussed enrollment, eligibility, and coverage, and provided a very simple sketch of how Medicare pays for services. We also briefly discussed how this information about Medicare influences the content of the data files used throughout the remainder of this book. In this chapter, we will build on those concepts and dive into each Medicare file we will be using for our example research programming project.[1] We will review the contents of each data set, including examples of information commonly pulled out of each file for research purposes. We will end with an example of how to use the datasets to identify services provided for a surgery and services provided for a visit to the emergency department.[2] Through these examples, we will see that the contents of each file are not always intuitive. Rather, the contents of each file are governed by how the services are billed and paid. We will learn to be cognizant of certain quirks, like the fact that not all services performed in an inpatient hospital appear in the inpatient claims data set. Looking forward, in Chapter 4, we will build on the information presented in this chapter by planning our programming project and, in Chapter 5, we will request the data described in this chapter. In subsequent chapters, we will actually use these data to address the research questions posed by our example project described in Chapter 1. For example, we will load and transform our claims files in Chapter 5, and calculate utilization of services in Chapter 7.

A Note on the Source and Structure of Our Claims Data

Before going into detail about the content of the Medicare administrative claims and enrollment data we will be working with, let's frame our discussion by identifying the source of our data and the structure of the files we will pretend to request, receive, and work with to complete our project.[3] It is important to introduce this conversation now because Medicare data can come from a variety of sources and each source stores and provides the data somewhat differently.

We will explore this topic in much more detail in Chapter 5, but for now it is important to know that, depending on our source of funding and the nature of our project, we could choose from at least four data sources:

CMS's data distribution contractor (sometimes referred to as the Research Data Distribution Center)

CMS's Data Extract System (DESY)

CMS's Virtual Research Data Center (VRDC) system

CMS's Integrated Data Repository (IDR)

For our purposes, we will assume that we will work with CMS's Research Data Assistance Center (ResDAC) and CMS's data distribution contractor to request and obtain our data.[4]

Medicare claims data created for research purposes are provided by CMS's data distribution contractor separately by claim type. In other words, separate sets of files are provided for final action non-institutional (Part B carrier and durable medical equipment) and institutional (outpatient, inpatient, skilled nursing facility, home health, and hospice) services. CMS further separates data for each claim type into separate files for base claim and claim line (in the case of the set of non-institutional data files) or revenue center (in the case of the set of institutional data files) detail, resulting in seven sets of files.[5]

The base claim detail (also known as header-level information) is basically summary information, including information that identifies the claim, the beneficiary who received care, the provider of service, the beginning and ending dates for the services billed on the claim, the diagnoses on the claim, and the total amount paid to the provider as reimbursement for the services performed. The claim line detail (sometimes generally referred to as line-level information) is a set of in-depth information about the specific services performed. For example, a line will contain a beginning and ending date of services, the patient's diagnoses (represented by diagnosis codes), the services performed (represented by procedure codes), payment amounts for the services, and the identifier of the provider that performed each service. In a SAS dataset, the claim is spread across multiple records with one record per service rendered.

In Chapter 5, we will transform these separate files to contain all claim- and line-level information in a single record. In other words, we will take the file with multiple records per claim and construct arrays on a single record to represent each of the line-items. This maneuver is not something you would do in your own research programming but, as we will discuss, the advantage is that we can work with a structure of claims data sets that can be created using data from any of the three sources. As such, this transformation "levels the playing field" and makes the remainder of the book accessible to a broader audience. In addition, such a structure just happens to be my preferred method of working with claims data because it affords the opportunity to see the entire paid claim in one record! Because such a transformation can consume valuable resources, more advanced readers should determine for themselves the most efficient way to structure their claims data.

Part B Carrier Claims Data

Carrier claims data contain claims information filed by non-institutional providers using the CMS-1500 claim form. When a Medicare beneficiary goes to the doctor for a checkup, the provider electronically submits a bill to Medicare Part B using the ANSI 837 5010 electronic format (or, using the CMS-1500 paper form). The CMS-1500 form contains details like the beneficiary and provider identifiers, dates of service, the procedures performed, and the patient's medical diagnosis. The submitted claim goes to the provider's regional Medicare Administrative Contractor (MAC) for adjudication, processing, and payment. The final action Medicare claims are then made available to the research community as the carrier claims data.

In my experience, users most readily associate this file with claims filed by individual physicians like internal medicine specialists or cardiologists. In other words, most users associate the carrier data set with claims filed for services provided in a doctor's office, like a checkup. However, the carrier data set also includes claims submitted by providers that you may not readily consider, like clinical social workers, chiropractors, ambulances, nurse practitioners, and physician assistants. In addition, the file also includes claims for services performed outside of a doctor's office, like ambulatory surgical centers, hospital outpatient departments, and hospital emergency departments.

Durable Medical Equipment (DME) Data

The DME file contains information on final action claims submitted by non-institutional DME providers using the CMS-1500 claim form. [6] These claims are processed by the provider's Medicare Administrative Contractor (MAC) for adjudication and payment. The final action Medicare claims are then made available to the research community as the DME claims data. Durable medical equipment includes wheelchairs and walkers, hospital beds, blood glucose monitors and related supplies, canes and crutches, splints, prosthetics, orthotics, respiratory devices like oxygen equipment and related supplies, and dialysis equipment and supplies.

DME is provided to any beneficiary with Part B insurance as long as it is medically necessary. Beneficiaries with Medicare Part B must have the equipment prescribed for use in their home by their doctor or "treating practitioner" (e.g., a physician assistant or nurse practitioner). A long term care facility can qualify as a beneficiary's home, but equipment is not covered if it is used in a hospital or a skilled nursing facility. For example, the hospital bed in an inpatient facility is included in the facility charge found in the Inpatient claims data.

Outpatient Claims Data

Outpatient claims data contain information on final action claims filed by institutional outpatient providers. [7] Most commonly users think of these providers as hospital outpatient departments. However, the data also includes the claims of other types of institutional outpatient providers like ambulatory care surgical centers, outpatient rehabilitation facilities, rural health clinics, and even community mental health centers.

Medicare Part B pays for many of the outpatient services a beneficiary receives in a hospital like the hospital charge for an emergency department service (as we will see below this does not include the doctor's charge), getting stitches or a cast, lab and X-ray services, outpatient surgery, observation required to determine if a beneficiary should be admitted for inpatient care, and even the administration of drugs if a beneficiary cannot self-administer the drug. For this reason, the claims in the outpatient data are sometimes referred to as "institutional Part B" claims. Looking forward to Chapter 8, we will see this in action when we do not use the outpatient data set when we sum total Part A costs.

A beneficiary is considered an outpatient if the beneficiary's doctor has not written the beneficiary an order to be admitted as an inpatient. These services are covered under Medicare Part B and are paid using the Outpatient Prospective Payment System (OPPS). The OPPS pays hospitals a predetermined payment rate (the rate differs by geographic region) to provide these services to Medicare beneficiaries.

Inpatient Claims Data

Inpatient claims data contain information on final action claims submitted by long-stay and short-stay inpatient hospitals for the reimbursement of their facility costs.[8] These claims are paid through Medicare Part A. Hospitals file claims for these services using the UB-92 form, also known as the CMS-1450 claim, or the electronic 837i format. It is important to make the distinction between facility costs and other costs for services provided in the hospital (again, more on this in the example below). Facility costs include things like room charges and even some drugs provided during a beneficiary's hospital stay. As we will discuss below, a doctor's visit or even a surgeon's services provided in the facility are billed using the CMS-1500 form and are therefore found in the carrier data.

A beneficiary is considered an inpatient starting the day the beneficiary's doctor formally admits the beneficiary to the hospital. The distinction between inpatient and outpatient status is very important because it influences how Medicare pays for the services received by the beneficiary. For example, as noted in Chapter 2, a beneficiary cannot be admitted to a skilled nursing facility unless the beneficiary has been discharged from a hospital within the last 30 days with an inpatient stay that lasted at least three consecutive days.

Skilled Nursing Facility (SNF) Claims Data

SNF claims data contain information on final action claims filed by SNF providers.[9] Like inpatient claims, SNF claims are billed using the CMS-1450 form or the 837i electronic format and paid through Medicare Part A. Skilled nursing facilities may be part of a nursing home or a hospital and provide skilled nursing and rehabilitative care through specialists such as registered nurses, physical therapists, occupational therapists, speech pathologists, and audiologists.

The purpose of this care is to treat, observe, and manage the conditions of beneficiaries leaving the hospital. The goal of skilled nursing care is to improve or maintain the beneficiary's current condition, as well as assist beneficiaries in maintaining their independence. Examples of skilled care include physical therapy and intravenous injections. Once admitted to a SNF (see the above section for more information on rules governing admission to a SNF), Medicare pays for SNF services for up to 100 days per spell of illness under specific circumstances. A spell of illness ends 60 days after discharge from a SNF, at which time a patient can be admitted to the hospital and then to a SNF for another 100 days.

SNFs are also paid on a perspective payment basis – resource use groups (RUGs). Each of the 66 RUGs has associated nursing and therapy weights that are applied to create daily base payment rates.

Home Health Claims Data

Home health claims data contain information on final action claims filed by Home Health Agency (HHA) providers.[10] Home health services are paid from Medicare Part A and Part B. Home health agencies are entities that provide skilled professional care in a beneficiary's home.

These agencies provide services to Medicare beneficiaries like occupational therapy, physical therapy, skilled nursing care, and even speech therapy and services like counseling that can help a beneficiary cope with the impacts of their illness on their mental health.

Medicare pays for home health services in units of 60-day episodes based on a prospective payment predetermined rate. Patients receiving five or more visits are assigned to 1 of 153 home health resource groups (HHRGs) based on clinical and functional status and service use which are adjusted for the geographical area where services are delivered.

Hospice Claims Data

The hospice claims data contain information on final action claims data submitted for hospice (also referred to as "end of life" or palliative care) services.[11] Hospice programs provide care for Medicare beneficiaries who are terminally ill. The focus of hospice care is on the comfort of the patient and not on healing or curing an illness.

A beneficiary qualifies for hospice services if the beneficiary has Medicare Part A and is terminally ill. A beneficiary is considered terminally ill if the beneficiary's doctor certifies that the beneficiary has less than six months to live, should the beneficiary's illness follow a normal progression.

Hospice services include physical care provided by doctors and nurses, care provided by hospice aides, occupational therapy, physical therapy, medical equipment and supplies, drugs, and even grief counseling and respite care (care provided in a facility designed to give family members a break from giving care) for the beneficiary's family members. All hospice services are covered as long as they are related to the beneficiary's terminal illness (services not related to the terminal illness are covered by other Medicare benefits). Many of these services may involve pain management, and services may take place in a facility or the patient's home.

Hospice care is insured in benefit periods, meaning that a Medicare beneficiary can get hospice care for two 90-day benefit periods and then an unlimited number of 60-day benefit periods. The doctor must certify that the beneficiary is terminally ill at the beginning of each period. Medicare beneficiaries have the right to stop hospice care at any time for any reason, and the right to change hospice providers once each benefit period.

Commonly Retained Elements in Administrative Claims Data

With about 47 million beneficiaries to account for, Medicare administrative claims files can be very large. Generally, the carrier claim file is the largest Medicare claims dataset used for research purposes. The large size makes sense because the file contains claims for the most common type of service, a visit to the doctor!

One very common way of easing the use of large datasets is to limit the files to contain only the key variables needed for your project. In Chapter 4, we will start the process for planning our project at which time we will begin to form a good idea of the variables we will need for the successful completion of our sample research programming project. At a minimum, when performing research, it is standard practice to keep the following variables in the administrative claims files:[12]

- The Medicare beneficiary's identifier:[13] Commonly referred to as the **HICAN** (usually pronounced "high-can") or **HIC**, this variable may not exist as its own data element in the Medicare research files we will receive. Although the data we will work with contains a ready-made, scrambled beneficiary identifier called a **BENE_ID**, you may have to create the **HIC** by concatenating the beneficiary's Claim Account Number (CAN) and the beneficiary's identification code (also known as the BIC). A beneficiary's identifier can change over time if a beneficiary's status changes (e.g., a beneficiary remarries). Therefore, it is important to have a master beneficiary identifier when looking at administrative claims and enrollment data over time.

- The provider identifier: A Medicare provider performed the services appearing on a claim, and each provider has a unique identifier that is reported on the claim record. For example, the National Provider Identifier (NPI) can be used to identify the provider that performed the services billed on the claim (known as the "performing" or "rendering" provider). The NPI is a 10-character identification number uniquely assigned to a Medicare provider. This identifier must be used by the provider for all financial transactions with Medicare. The NPI is known as an intelligence-free identification number, meaning that it does not contain any information about the provider, like specialty or Medicare region. We will see that the provider identifier for hospitals found in inpatient claims data does contain embedded intelligence.
- Codes identifying services: The services performed by the Medicare provider can be identified using procedure codes represented by Healthcare Common Procedure Coding System (HCPCS) codes. HCPCS Level I codes, also known as Current Procedural Terminology (CPT) codes, are five digit character codes describing medical services, and are a copyrighted coding schema of the American Medical Association (AMA). HCPCS Level II codes are five character alphanumeric codes defining services not described by the Level I codes.14 In other cases, revenue center codes and ICD-9-CM procedure codes (more on this below) can be used to identify services Revenue center codes correspond to cost center units (or divisions) within a hospital, like emergency departments.
- Codes identifying the beneficiary's medical condition: The International Classification of Diseases, Ninth Revision, Clinical Modification (ICD-9-CM) is the United States' method of assigning medical diagnoses to patients (like diabetes, COPD, or prostate cancer). As of the writing of this text, in October 2014, ICD-9-CM will be replaced by ICD-10-CM.15 ICD-9-CM can also be used to identify procedures in inpatient, skilled nursing facility, and outpatient claims.
- Dates of service: This refers to the dates of service of the procedures performed by the provider. This can include specific dates of service for specific procedures (from detail service lines), or admission and discharge dates (header information) from a stay in an institution (like a hospital).
- Payment information: The total cost of a claim includes the amount paid by Medicare, as well as deductible and coinsurance payments. We are interested in the total cost to Medicare for the services on a claim, so we will utilize only the Medicare payment variables (without using information such as deductibles and coinsurance payments).

This list is by no means exhaustive. In some cases, investigators need to look at payment information like deductibles and coinsurance, provider identifiers like Tax Identification Numbers or CMS Certification Numbers, and information on the provider's medical specialty. In your work, you may need to utilize information in inpatient claims data such as the source of admission to the hospital, the discharge status or discharge destination, and a Diagnosis Present on Admission (POA) indicator (a field that identifies whether a diagnosis was already present upon admission to the hospital). In addition, some investigators may wish to review the Medicare Severity-Diagnosis Related Group (MS-DRG), a patient classification system comprised of a set of codes developed for Medicare. MS-DRGs use information from a beneficiary's paid hospital claims (like age and medical diagnosis) to classify hospital services provided to patients. Finally, you may wish to review the level of hospice care provided (e.g., routine home care, inpatient respite care, and continuous home care).

As you can imagine, even this small set of data elements can be used to paint a very detailed story about the nature of a particular claim, like a visit to the doctor for a checkup. When combined with other final action claims, this information can be used to draw broader conclusions about a particular beneficiary or provider.

Master Beneficiary Summary File

We have reviewed seven files containing paid claims data, but our project also requires information about the beneficiaries we are studying. When I started using Medicare claims data, the most common source of data for beneficiary demographic information was the Denominator file. The name "Denominator file" may seem odd, but it is appropriate. In many applications of beneficiary demographic information, a calculated number (say, the number of visits to a doctor's office) is divided by a count of beneficiaries (perhaps a sample population). For example, if we calculated the total utilization of emergency department (ED) services, we may wish to get the per capita utilization rate by dividing by the total number of beneficiaries in our sample. In this equation for per capita utilization of ED services, the count of sample beneficiaries is the denominator! Information from the demographic file is often used as the source of denominator in these kinds of equations. Therefore, the file containing this beneficiary information for the full Medicare population for a given year is named the Denominator file.

In March 2010, my beloved Denominator file was integrated into something called the Master Beneficiary Summary File (MBSF).[16] The MBSF contains demographic information for beneficiaries who were alive and eligible for Medicare at any time during the calendar year for which the file was created. This information includes the beneficiary's identifier, sex, date of birth, and date of death (if applicable). In addition, it includes geographic information like the beneficiary's state and county codes. Perhaps most importantly, it includes information pertaining to the beneficiary's enrollment in the Medicare program, like data elements that can be used to determine the beneficiary's enrollment status in Medicare Parts A, B, C, and D.

Provider Data

For our example research programming project, we assume that we have been given a list of the providers in our study population (along with their associated beneficiaries), and we do not require information on those providers above and beyond what is contained in the paid claims data. For example, the Part B carrier file contains information such as a performing provider's NPI, provider type (e.g., group, solo practitioner, independent lab, and institutional provider), specialty, state, zip code, and specialty code.

CMS also provides descriptive data on providers through the National Plan and Provider Enumeration System (NPPES, pronounced "in-pez") file. The NPPES file is a publicly-available dataset that can be downloaded from CMS's website. The file contains data elements like the provider's NPI, information on whether the provider is an individual or an organization (like an ambulatory care center), the Health

Care Provider Taxonomy Code (HPTC, which is, in very simple terms, a code that describes the provider's specialty), name, business name and address, contact information, and license number.[17]

Example: Identification of Emergency Department (ED) Utilization

Let's say you are on a game show called Medicare Millionaire, where you have to answer research questions using Medicare claims and enrollment files. Personally, I would tune in faithfully every night, but I have a feeling that I may be one of the only viewers (after all, it is a small group of people who do this work)! Imagine that you have successfully answered several questions and now have a chance to win a large cash prize based on your answer to the following question: Which of the claims data files are used for the identification of ED utilization? If you already used your lifeline to call a friend, you would need to take your best guess. What would that guess be? Without any knowledge other than what we have learned in this chapter, my guess would be the inpatient claims data. Considering that emergency rooms are typically found in inpatient hospitals, this would be a good choice! Unfortunately, the answer is only partially true!

Complete information on emergency department-related information can be found in no less than three sets of claims data: inpatient data, outpatient data, and Part B carrier data![18] Although we will write code to calculate ED utilization for our example research programming project, discussing how to identify ED utilization at this time is enlightening. There are several factors that make identification of ED services complex. First, there is the issue of whether or not the beneficiary was admitted to the facility for an inpatient stay. Second, there is the issue of identifying what is often referred to as the "professional component" of the service (described below).

If a beneficiary is admitted to the hospital following the ED visit, the information pertaining to the ED visit can be found in the inpatient data set. These ED services are identified using revenue center codes 0450-0459 and 0981. The inpatient claims data contain diagnostic information gleaned from the emergency visit. On the other hand, if the beneficiary is seen in an ED and not subsequently admitted to the hospital, the claim for the ED visit can be found in the outpatient data! These claims can be identified using the same revenue center codes used for identification of ED visits in the inpatient data.

The professional component of a service can simply mean the portion of the service provided by a physician. For example, when a lab test is performed, the "professional component" is a provider's (typically a pathologist or other physician) interpretation of the test and the "technical component" is the administration of the test itself. In the case of ED services, the professional component will appear in the Part B carrier data because it is billed using the CMS-1500 claim form and paid through the beneficiary's Medicare Part B coverage.

Again, we will construct SAS code to identify ED utilization in Chapter 7. For the time being, this example illustrates the complexity of using Medicare data to identify utilization of some common services. Just in case you are not convinced of the potential pitfalls inherent in using Medicare data for research purposes, let's discuss another example, the identification of surgical services.

Example: Identification of Surgical Services

Let's say that you made it through to the bonus round of Medicare Millionaire. Maybe you are starting to sweat under the hot lights as the host lines up a really hard question. After a short commercial break to build the anticipation, the host throws a real hardball question at you: "How do you identify surgical services in Medicare claims?" Wow! Hopefully, you smile and answer the question knowingly because the previous night you brushed up for the game show by reading the following information in this chapter!

The identification of surgical services in Medicare claims data requires the use of inpatient, outpatient, and Part B carrier data sets. If you are interested in post-acute care, you can even add the home health, hospice, and SNF data sets to the list! Let's start with tracking down all of the information related to the surgery by looking in the most intuitive location first: the inpatient data set. After all, when most folks think of surgery, they picture an operating room in an inpatient hospital setting. Surgeries performed in an inpatient setting are typically part of an inpatient stay and found in the inpatient claims file. That said, surgeries are commonly performed in the outpatient department of hospitals and these claims appear in the outpatient data set. Finally, surgeries can also be performed at Ambulatory Surgical Centers as well, and these services appear in the outpatient and Part B carrier data sets. The facility charges for the surgeries provided in any of these settings can be found in the data mentioned above. However, the story doesn't end there! There is also the matter of the surgeon's professional services and, lest we forget (or assume the beneficiary was asked to bite down on a stick), the anesthesiologist's services. These services appear in the Part B carrier data set.

Chapter Summary

This chapter provided an introduction to the Medicare claims and enrollment data commonly used in research programming efforts. We learned:

- Data exist for institutional and non-institutional claims paid by Medicare Part A and Part B.
- There are seven types of paid claims (not including Part D information): Part B carrier, DME, outpatient, inpatient, SNF, home health, and hospice.
- For instructional purposes, we will request our claims and enrollment data from CMS's data distribution contractor. Non-institutional data arrive in separate files for Part B carrier base claim, Part B carrier claim line, DME base claim, and DME claim line detail. Institutional data arrive in separate files for base claim and revenue center information (claim line detail) for each claim type. In Chapter 5, we combine these base claim and claim line (or revenue center) information into single records by claim type.
- At a minimum, the following data elements are commonly required for research programming with claims data:
    - Information identifying the beneficiary identifier, such as a **BENE_ID** or **HIC**.
    - Information identifying the provider of the service, such as the NPI.

    - Information describing when the services occurred (start and end dates, or admission and discharge dates).
    - Diagnostic codes describing the medical condition of the patient, and procedure codes or revenue center codes describing services provided to a beneficiary.
    - Information on the amount paid by Medicare for the services on the claim.
- Enrollment data exist for Medicare beneficiaries. Medicare beneficiary enrollment data include information on beneficiary date of birth, date of death, residence, and reason for entitlement and FFS enrollment status.
- Using Medicare claims is not intuitive. Identifying all components of ED or surgical services involves the use of multiple claims files.

---

[1] The information presented in this chapter is derived from resources like the Medicare coverage publications available at www.medicare.gov (see, for example, http://www.medicare.gov/publications/pubs/pdf/10153.pdf). These manuals are wonderful sources of information. It also draws on information available at www.ccwdata.org, www.resdac.org, and www.medpac.gov.

[2] We will not write SAS code in this chapter because we want to stay focused on learning about the Medicare program and Medicare data. In Chapter 7, we will construct SAS code for identifying emergency department services. Writing SAS code to identify surgical services can be rather advanced, and is outside the scope of this text.

[3] The discussion of acquisition of source data is for instructional purposes. The data we use in this book are fake, and subject to the disclaimer presented in Chapter 1. In other words, although we pretend to go through the process of requesting data from CMS, we are simply describing how the process would occur, and we did not actually procure the data used in this book through that process.

[4] Although we are pretending that we are receiving data from CMS's data distribution contractor, we are actually using CMS's Data Entrepreneurs' Synthetic Public Use File (DE-SynPUF), which is available in the public domain. Please see the disclaimer in Chapter 1 for more information.

[5] Data from DESY are not received in this same fashion, nor would the output of a query of the IDR.

[6] See CMS's booklet called Medicare Coverage of Durable Medical Equipment and Other Devices, available at http://www.medicare.gov/Pubs/pdf/11045.pdf, for a more comprehensive treatment of the information provided in this section.

[7] See CMS's pamphlet called Are You a Hospital Inpatient or Outpatient, available at http://www.medicare.gov/Pubs/pdf/11435.pdf, for more information on Medicare's outpatient benefits.

[8] It is important to note that the MedPAR file also provides information related to inpatient (and SNF) claims. This file is summarized at the stay level and may provide the right level of detail for many research programming exercises.

[9] See CMS's booklet called Medicare Coverage of Skilled Nursing Facility Care, available at http://www.medicare.gov/Pubs/pdf/10153.pdf, for a more comprehensive treatment of the information provided in this section.

---

[10] See CMS's booklet called Medicare and Home Health Care, available at http://www.medicare.gov/Pubs/pdf/10969.pdf, for a more comprehensive treatment of the information provided in this section.

[11] See CMS's booklet called Medicare Hospice Benefits, available at http://www.medicare.gov/Pubs/pdf/02154.pdf, for a more comprehensive treatment of the information provided in this section.

[12] More information on the variable names is provided in later chapters. Readers should refer to the data dictionaries available at www.resdac.org or www.ccwdata.org for more information on the variables available.

[13] Depending on the nature of your data request, you may not be provided with the **HIC** (because it identifies a beneficiary and can be a Social Security Number). Rather, you may be provided with a de-identified beneficiary identifier.

[14] See CMS's HCPCS-General Information website available at http://www.cms.gov/Medicare/Coding/MedHCPCSGenInfo/index.html?redirect=/medhcpcsgeninfo/.

[15] For more information on ICD-10, visit the CDC's International Classification of Diseases, Tenth Revision, Clinical Modification (ICD-10-CM) page at http://www.cdc.gov/nchs/icd/icd10cm.htm.

[16] More information on the MBSF is available on ResDAC's website at http://www.resdac.org/cms-data/files/mbsf.

[17] See CMS's NPPES Data Dissemination-Readme file available at http://www.cms.gov/Regulations-and-Guidance/HIPAA-Administrative-Simplification/NationalProvIdentStand/Downloads/Data_Dissemination_File-Readme.pdf.

[18] For more information, see ResDAC's article at http://www.resdac.org/resconnect/articles/144.

Chapter 4: Planning the Research Programming Project

Introduction and Goals of This Chapter

Our goal in this chapter is to plan our research programming project. As you will see, the planning process is a fundamental part of any research programming effort. In my experience, aside from a misunderstanding of data, the lack of a methodical and comprehensive programming plan is the most common reason that research programming efforts struggle or even fail.

I bet I know what you are thinking right now: "Are you kidding me? We are talking about *research* programming! I mean, how can you plan something that no one has done before? By its very nature, research means adapting to new information and changing approaches to answering questions! Planning is just a waste of time! Besides, I want to get cracking and write some code!" I have to admit that it is very tempting to skip the planning process and just dive right in to the work! After all, for many programmers, writing out a methodical plan is right up there with major dental work! It is not something most people consider particularly fun, especially when we could be combing through data and answering research questions! However, be forewarned: Failure to plan properly will greatly increase the likelihood of rework, costing valuable time and money (and causing unneeded stress and anxiety). For example, can you imagine coding for three months only to realize you forgot to request an important variable? What if you had to put the project on hold while waiting for data to be re-extracted and delivered?

The cornerstone of this chapter is that planning a research programming project is not only possible and necessary, but extremely helpful. Although a formal project plan and documentation may seem like overkill or unnecessarily rigid, it is important to remember that even a small research programming project can become very complicated, use many datasets, process large files, generate a tremendous amount of code, and produce numerous output files.

Viewing a research programming project as the construction of an information system and using traditional information systems approaches to planning mitigates the risk of error and produces a higher quality product. Besides, although no two research programming projects are exactly alike, many similarities exist that make planning a little easier. For example, most every research programming project needs to acquire data, create files for analysis, calculate and summarize results, and generate output. In addition, many research programming efforts must modify code based on the iterative feedback of the research process. Finally, many projects reference code and, importantly, the reasoning behind methodological and coding decisions, many months or even years after the completion of a particular task or the project as a whole (you would be surprised at the speed with which important details and reasoning leave your brain!).

Many planning tools exist and many books cover the science of planning a programming project much more comprehensively than this chapter![1] The tools you choose must meet the specific needs of your project. In my experience, the tools discussed in this book are applicable to the vast majority of research programming efforts. However, there are projects for which you should consider alternative methods. For example, projects that produce reports on a quarterly basis may need a more rigorous and formalized approach that includes separate development and production phases and change request documents. Very short-term projects that will produce a single, simple deliverable may not require much of a formal planning structure.

In this chapter, I recommend an approach to planning a medium or large research programming project, starting with the development of an overall project plan based on the Systems Development Life Cycle (SDLC), an organized approach to planning and executing a programming project. This plan includes instructions for designing quality assurance (QA) procedures, developing a requirements document that defines the objectives of the project at a high level, and creating a set of technical specifications that detail the programming steps required to meet the project objectives.

By the end of this chapter, we will develop our overall project plan, as well as our high-level requirements and QA plan. We introduce the concept of technical specifications, but we will tackle the development of detailed technical specifications in subsequent chapters as we write code.

Our Project's Plan: The SDLC

There are many ways to approach the development of a plan for research programming work. Depending on the complexity and duration of your project, it may suffice to list all of the known tasks starting with requesting data and ending with the production and archiving of output and SAS code. For projects that are more complex, like the research programming effort covered in this book, I prefer to execute the planning process within the framework of the SDLC.

The SDLC is a systematic approach to creating an information system. The SDLC divides the creation of a system into phases, including the development of requirements, the design of the system based on the requirements, the creation and testing of code, implementation of the code, and disposition of the system. While these phases may sound foreign, they are really quite simple when expressed in terms that are more familiar. The SDLC brings order to managing a project by asking those working on the project to:

- Fully develop, define, and document the research questions and the approach to answering those questions
- Develop code that adheres to the approach to answering the questions
- Test, quality assure, execute, document that code, and provide output
- Close out the project (including the destruction of protected, identifiable data as required by CMS)

You can choose from many different SDLC methodologies. Some methodologies, like agile development, facilitate changes during the development of code. Other methodologies, like waterfall, rely on the creation of a more comprehensive plan from the start and, therefore, are less conducive to a project that relies on an iterative process.

Many authors have written books on the SDLC, and a complete study is out of the scope of this text. For our purposes, we will not focus on choosing an SDLC methodology because the basic steps of any SDLC methodology are a sufficient framework for our project plan. The basic steps of the SDLC, as adapted for our research programming purposes, are:

1. Initiation, Planning, and Design
    a. Develop and finalize high-level project requirements
    b. Develop a plan for quality assurance
    c. Develop and finalize a project flowchart
    d. Develop and finalize initial research programming specifications
    e. Develop quality assured programming specifications and system flow diagram
    f. Request data based on programming specifications and systems design
2. Develop and Test Code
    a. Develop code according to plan and design
    b. Test and quality assure code
3. Implementation
    a. Execute code in batch and retain logs, lists, and milestone datasets

4. Disposition
   a. Document code
   b. Archive final code, logs, lists, and data (if applicable and according to contractual arrangements)
   c. Destroy data as applicable

For the remainder of this chapter we will focus on fleshing out Step 1. More specifically, we will flesh out high-level requirements and the quality assurance plan for our project (Step 1.a. and Step 1.b.), and discuss the programming technical specifications (Step 1.c. and Step 1.d). However, as discussed in the introduction, we will not actually write the technical specifications until we tackle coding in subsequent chapters. The discussion of technical specifications in subsequent chapters is really for instructional purposes; in an ideal world, we would draft technical specifications prior to launching our programming effort and simply tweak them as we work through our research. Therefore, Chapters 5-8 focus on Step 1.d. through Step 1.f. (Initiation, Planning and Design), as well as Step 2 (Develop and Test Code). In fact, Chapter 5 through Chapter 9 (and some of Chapter 10) essentially read like very detailed technical specifications that contain a lot of instructional information! Chapter 10 includes information on Step 3 and Step 4.

Our Project's High-Level Requirements

The purpose of high-level requirements is to state the objectives of the project in very broad terms. These requirements assist us in getting a handle on the goals of the project and staying on course throughout the life of the project. Typically, requirements list the research questions, source data needs, output requirements, and timeframes. The high-level requirements for our research programming project are:

- We will write programs to answer the research questions described in Chapter 1.
- We will create descriptive output to answer the research questions posed in Chapter 1.
- We will use Medicare claims and enrollment data for calendar year 2010. Specifically, we will request enrollment data and all Medicare Part A and Part B claims data for all beneficiaries seen by the providers in our study population for that calendar year.
- We will use a desktop PC with ample storage and BASE SAS to perform all data work for our project.
- We will complete our project objectives in three months.
- We will destroy all data in compliance with contractual agreements upon successful completion of the project.

Our Project's Quality Assurance Plan

The purpose of a quality assurance (QA) plan is to explicitly state the project's QA objectives and determine a plan of attack for meeting those objectives. Our objective is to develop code that accurately reflects the objectives outlined in the programming specifications. To this end, we will adhere to the following QA rules:

- We will QA our programming specifications.
- We will perform unit testing (testing individual sections of code) and debugging.
- We will employ technical peer reviews of our code.
- We will test the integration of the system as a whole.
- We will investigate the validity of our output.
- We will batch submit our code in production and clearly label and save our logs and lists. In addition, we will erase output before resubmitting (this avoids accidentally using data from prior submissions of code).

The reader will need to trust me that the development and integration of our code is unit tested and debugged and the requirements and code were reviewed by peers. The reader will also need to take me at my word that I batch submitted all code, as well as clearly labeled copies of logs and lists and adhered to sound data management practices.

Our Project's Flowchart

This section discusses the creation of a flowchart of the steps involved in executing our research programming project. Flowcharts and diagrams help the programmer discover and work through design issues both within programs and between programs that, when integrated, comprise a system. In addition, they serve as a very useful form of documentation, summarizing many pages of complex processes in a couple of pictures. Before we go too far down the road of creating our flowchart, it is helpful to discuss terminology and distinguish between common types of flowcharts and diagrams. Terms like flowchart, flow diagram, process flowchart, and process map are often used interchangeably. Because this is not a book about flowcharts and diagrams, we will keep things simple and use the term "flowchart" when discussing any picture that describes how processes and controls work together as a project, a program, or a system. We will use the term "diagram" when we describe the flow of data. More specifically, in this book:

- Project flowcharts focus on the workflow of tasks for a project, or within a specific task in a project. We will create a workflow diagram for our project in this section.
- Program flowcharts describe the flow of programs within a system (a set of programs), while others describe the flow of processing steps within a single program.[2]
- Data flow diagrams describe the flow of data within or between programs.

The below project flowchart, Figure 4.1, focuses on the flow of tasks that we must accomplish to complete our research programming project. The purpose of our flowcharting exercise is to document the steps involved in executing our research programming project, describe how steps may interact with each other, and assist in planning our work. It is the equivalent of a blueprint for building a house and helps us plot out each step and identify the most efficient arrangement of our activities.

Figure 4.1: Research Programming Flowchart

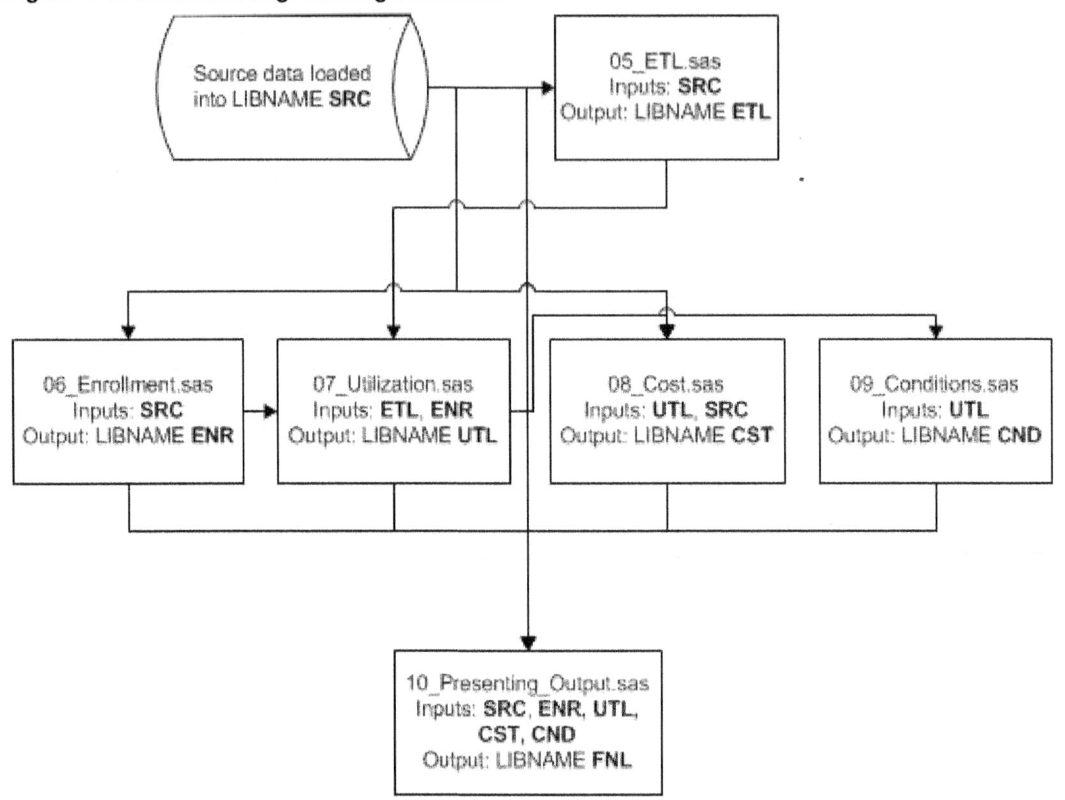

This flowchart pictorially represents our programming plan as follows:

- As we will discuss in Chapter 5 (and mentioned briefly in Chapter 1), our source data are comprised of extracted claims and enrollment SAS data sets, as well as a finder file of providers and associated beneficiary identifiers. In addition, as we will discuss in Chapter 6, we downloaded a geographic code file from the internet into our source data library.
- The program **05_ETL.sas**, developed in Chapter 5, transforms our source claims SAS data files. The output is used in our work in Chapter 7 (to delimit our claims data using the continuous enrollment information derived in Chapter 6).

- The program **06_Enrollment.sas**, developed in Chapter 6, uses the enrollment source SAS data set to create a continuous enrollment data set. It also uses the geographic codes file loaded into a SAS data set to add state and county names to a data set of continuously enrolled beneficiaries. This output is used in Chapters 7 and 10 for demographic information and to delimit our claims data, keeping only those claims for continuously enrolled beneficiaries.
- The program **07_Utilization.sas**, developed in Chapter 7, uses the source claims SAS data sets transformed in the program **05_ETL.sas** as well as the continuous enrollment data created in the program **06_Enrollment.sas** to create claims files delimited to contain only claims for continuously enrolled beneficiaries. It also creates inpatient claims data delimited to acute short stay hospitals. These claims are used to calculate utilization, cost, and quality measurements in Chapters 7, 8, and 9, respectively. In addition, utilization summary files created in the program **07_Utilization.sas** are used in Chapter 10.
- The program **08_Cost.sas**, developed in Chapter 8, uses the claims data created in the program **07_Utilization.sas** to compute measures of cost to Medicare. It also uses the raw carrier source data set to perform an exercise in working with a data set comprised of one record per claim line. Cost summary files created in the program **08_Cost.sas** are used in Chapter 10.
- The program **09_Conditions.sas**, developed in Chapter 9, uses the claims data created in the program **07_Utilization.sas** (as well as the E&M utilization summary created in that same program) to identify beneficiaries with diabetes or COPD, and to compute certain measurements for those beneficiaries. Summary files created in the program **09_Conditions.sas** are used in Chapter 10.
- The program **10_Presenting_Output.sas**, developed in Chapter 10, uses summary files created in the programs **07_Utilization.sas, 08_Cost.sas**, and **09_Conditions.sas**, as well as enrollment data created in **06_Enrollment.sas**, and our "finder file" of providers and associated beneficiaries.

Our Project's Data Storage and LIBNAMES

Before we begin any programming, it is worth drawing on the above discussion to determine the SAS libraries we will use for storing our example project's data sets. We will use the following LIBNAME statements for the remainder of our work. Our data storage plan involves separating input and output data sets logically by task (or, in our case, chapter). The storage path may differ for each reader based on the organization of storage on your particular system.

- **libname SRC 'C:\DATA\00_Source_Data'**: This library contains all data received from CMS's data distribution contractor, loaded into SAS data sets using code that came with our data (which will not be displayed in this book). The library also contains any other source data files, including the finder file SAS data set received from CMS and geographic codes file. I recommend making a backup copy of this source data folder if possible—as long as we are acting in accordance with any contractual arrangements, and we remember to delete it as part of the disposition of our project (see Chapter 10).
- **libname ETL 'C:\ DATA\05_ETL'**: This library contains the "extracted, transformed, and loaded" data. For example, this folder contains the claims data modified to contain one claim

per record (as briefly discussed in Chapter 3 and covered in more detail in Chapter 5) for use throughout the remainder of our project.

- **libname ENR 'C:\DATA\06_Enrollment'**: This library contains the output (and milestone data) from our work on enrollment. For example, the data set containing enrollment and state and county name information for beneficiaries identified as continuously enrolled throughout 2010 (ENR.CONTENR_2010_FNL) is stored in this library.
- **libname UTL 'C:\DATA\ 07_Utilization'**: This library contains the output (and milestone data) from our work measuring utilization.
- **libname CST 'C:\DATA\08_Cost'**: This library contains the output (and milestone data) from our work measuring cost.
- **libname CND 'C:\DATA\09_Conditions'**: This library contains the output (and milestone data) from our work studying chronic conditions and simple quality measurements.
- **libname FNL 'C:\DATA\10_Results'**: This library contains the output (and milestone data) from our work presenting the results of our example programming project.

Note that the folders are numbered so they can be sorted to appear in the order in which the code is executed.

Our Research Programming Technical Specifications

As stated above, we will tackle the creation of detailed technical specifications in Chapters 5 through 10. However, it is worth pausing to introduce the concept of technical specifications in advance of those chapters. The purpose of programming specifications is to guide the development of SAS code by providing written instructions on what the SAS code must do to create the required output. In addition, specifications document the reasoning behind important methodological and coding decisions. Specifications come in many forms and describe what needs to be done at many different levels; some specifications state the project objectives ("what" must be done) at a very high level, while other types of specifications provide a lot of detail on "how" to do what needs to be done.

Research programming specifications are different from other kinds of programming specifications in that they contain less upfront detail and more room for iterative development and discovery. Typically, good research programming specifications state the objectives of the project (by clearly describing the research questions intended to be answered) and provide information on the inputs required (including the datasets and variables needed), the calculations to be performed, and the required output. Therefore, team members that understand the research questions and the data must write the specifications. Even if you are a one-person team who is formulating the research questions and answering them by writing and executing SAS code, it is worth creating specifications because they help identify all of the processing steps upfront. At the least, it is a best practice to use programming specifications to guide any code walkthrough or technical review of our code.

Construction and carpentry provide many excellent metaphors for programming. I am fond of home improvement projects (and it is a good thing I am, because there is no shortage of such projects around my house, which was built in 1855!). I often find myself thinking about how building a table, installing a chair rail, or patching up an old wall made of field stone is like building a system of code. For example, we can consider plans based on the SDLC to be architectural blueprints. Similarly, technical specifications are the programmer's equivalent to the carpenter's mantra of "measure twice and cut once." Technical specifications help you think through what data you will need, what variables you will use, and any issues related to data processing and storage. If you cut a piece of crown molding to an incorrect length, it can result in the need to purchase (or worse, rebuild) replacement crown molding and perform the measuring and cutting again. In the same way, if you fail to identify and think through all of the major steps in your programming effort, it can result in moving forward with data extraction and coding only to later realize that you left something out. For example, planning how you will identify beneficiaries who are continuously enrolled in Medicare Fee-for-Service helps you to formulate your study population. Repeating the data extraction for the study population to compensate for a mistake can result in a lot of lost time! In the end, even the best plans will not execute exactly as planned (it is research, after all), so it is important to document changes that happen along the way.

Chapter Summary

In this chapter, we planned our research programming project. The planning process is a fundamental part of any research programming effort. The fact that we are performing research and expect to encounter many unknowns is not an excuse for circumventing the planning process. Rather, the existence of unknown information in our project is a very good reason to devote a lot of time to formulating our project plan. Indeed, the lack of appropriate planning is perhaps the major reason for the failure of research programming projects. To this end, we discussed the following:

- Our recommended approach to planning a medium or large research programming project involves starting with the development of an overall project plan based on the Systems Development Life Cycle (SDLC).
- A plan based on the SDLC includes instructions for quality assurance (QA), a requirements document defining the objectives of the project at a high level, and a set of technical specifications that describe in detail the programming steps required to meet the project objectives.
- The basic steps of the SDLC are initiation, planning and design, developing and testing code, implementation, and project disposition.
- High-level requirements state the objectives of the project in very broad terms, and assist us in identifying the goals of the project and staying focused throughout the life of the project. Typically, requirements list the research questions, source data needs, output requirements, and timeframes. We developed high-level requirements for our research programming project.
- A QA plan explicitly states the project's QA objectives and determines a plan of attack for meeting those objectives. We developed a QA plan for our research programming project.
- Flowcharts and diagrams help the programmer discover and work through design issues both within programs and between programs that comprise a system. In addition, they serve as a very useful form of documentation, summarizing many pages of complex processes in a couple of

pictures. We developed a flowchart of the major tasks involved in our research programming project.

- Technical specifications guide the development of SAS code by providing written instructions on what the SAS code must do to create the required output. In addition, specifications document the reasoning behind important methodological and coding decisions. Good research programming specifications state the objectives of the project and provide information on the inputs required, the calculations to be performed, and the required output. We will develop our project's technical specifications in Chapter 5 through Chapter 10.
- We laid out the data storage strategy for our programming work, numbering the folders so they can be sorted in the order in which the code is executed.

---

[1] See, for example, McConnell, Steve. 2004. *Code Complete, Second Edition.* Washington: Microsoft Press.

[2] It would be best to have detailed technical specifications in hand when creating the flowchart for a program or system of programs. We will not create flowcharts for our programs or data flow diagrams in this book, but I do recommend it for planning and documentation purposes.

Chapter 5: Request, Receive, Load, and Transform Data

Introduction

The purpose of this chapter is to discuss the process for requesting, receiving, loading, and transforming Medicare administrative claims and enrollment data. As discussed in Chapter 1, the online companion to this book is at http://support.sas.com/publishing/authors/gillingham.html. Here, you will find information on creating dummy source data, the code in this and subsequent chapters, as well as answers to the exercises in this book. I expect you to visit the book's website, create your own dummy source data, and run the code yourself.

Requesting Medicare Administrative Claims and Enrollment Data[1]

From scheduling and budget perspectives, it is very important to keep in mind that a data request can take a lot of time and cost a lot of money to fulfill (although there are ways to defray this cost for researchers[2]). Therefore, it is important to plan as much as possible prior to initiating our data request so that we minimize the likelihood of needing to request a re-extraction because of missing files or variables. Because research programming involves an iterative investigatory process, we will never be perfectly certain that we have identified every data need. However, I have found that executing the kind of planning discussed in Chapter 4 not only minimizes the risk of rework, but also improves the project's ability to absorb rework if necessary. Therefore, now that we have determined our research programming project's plan and spent some time thinking about the high-level requirements, we know enough about our project to request data. More to the point, we know the research questions we need to answer and the data we will use to answer those questions. We will use this information to guide our data request. Indeed, as we will see, the successful completion of our programming project requires requesting and using personally identifiable beneficiary-level claims and enrollment data.

Sources of Medicare Claims and Enrollment Data

The data we need to complete our research programming project are not generally available to the public. We must obtain the data from CMS, a process that requires requesting, receiving, and loading the data. Recall that we touched on the concept of Medicare administrative data sources in Chapter 3 in order to frame the discussion of the contents and structure of the Medicare claims and enrollment data we will use to complete our research programming project. We will now review and expand on that discussion.

There are several routes to procuring Medicare data for research purposes.[3] Depending on our source of funding, some routes may not be available to us. In the extreme case, employees of CMS have access to data from a variety of internal sources, including CMS's Data Extract System (DESY), CMS's Virtual Research Data Center (VRDC), and the Integrated Data Repository (IDR). Researchers not employed by CMS, including federal contractors and academics, may also have access to DESY, the VRDC, and the IDR but, depending on the source of funding for their project, use of these sources may not be permissible. Instead, contractors and academics may need to access Medicare claims and enrollment data by working with CMS's Research Data Assistance Center (ResDAC) and CMS's data distribution contractor.

So where are we going to get the data for our research programming project? From an instructional perspective, this is a tough question. Indeed, there are differences involved in using each of the sources that would be useful to know and that can influence what programming techniques we discuss in this text. Here are some examples:

- Submitting a request through ResDAC to be filled by CMS's data distribution contractor involves planning for certain paperwork and delivery timeframes. Using DESY, the VRDC, or the IDR also involves certain planning activities, such as acquiring a user account with CMS.

- Using the IDR requires skill in querying a relational database, as well as more specialized knowledge to take advantage of attributes like TeraData indexing.
- Each data source may deliver the data in a different format. For example, data from DESY is typically delivered on a mainframe environment as flat text files (access to CMS's mainframe environment, as well as knowledge of how to load the flat file, is required to work with these files).

With this in mind, our instructional plan is based on three key assumptions:

1. We will request our source data through ResDAC and CMS's data distribution contractor because it is probably the most common route to Medicare data traveled by the research community. Besides, as we just discussed, other routes like DESY and the IDR require specialized skills that fall outside the scope of this text, and access to these routes are uncommon for researchers. Therefore, in this chapter, we will review how to request and receive identifiable Medicare claims and enrollment data through ResDAC and CMS's data distribution contractor; we will no longer focus on other data sources, like DESY, the VRDC, or the IDR.
2. Our source data will arrive from CMS's data distribution contractor as files that we must read into SAS data sets. However, we will not spend time learning to load the flat files into SAS data sets because CMS's data distribution contractor provides SAS programs that load the files into SAS data sets. We start our programming work by using SAS data sets, assuming we have already gone through the process of loading the files we received from CMS's data distribution contractor into these SAS data sets.
3. Our source data will arrive in separate files by claim type, with each claim type separated into files for claim header (called "base claim" files) and claim line information. We will create files that contain both base claim and claim line data in a single record (thus halving the number of claims data sets we work with) by combining the base claim and line files.[4] Creating and using claim-level files for instructional purposes makes it easier to understand the concept of a claim as a bill for services rendered because all of the information on a claim is contained in a single record. Well, at least it has always been easier for me to understand it and teach it that way! Most importantly, it provides an achievable, common starting point irrespective of the source of our data, and thereby makes our discussions accessible to a broader audience.[5]

With our instructional plan set, let's now discuss a little more about data available through CMS's data distribution contractor, describe the process of requesting the data needed for our example research programming project, and write code to create the claims and enrollment data used throughout the remainder of this text.

Data Available from CMS

Our example research programming project calls for the use of personally identifiable (we define "personally identifiable" below) claims and enrollment data files. However, it is worthwhile to pause for a moment and discuss other data available to the research community from CMS. A complete review of all Medicare data available to the research community is outside the scope of this book, but I would hate to leave the reader without a general understanding of the Medicare data CMS offers to the research community. Who knows, you may need something like non-identifiable data for future projects, so we may as well spend a little time obtaining a broader view of the Medicare research data landscape!

CMS offers a dizzying array of datasets for research purposes. The most straightforward method of understanding the Medicare data available to the research community is to group the files into the following three categories:

1. **Research Identifiable Files (RIFs)**: Claims and enrollment files containing personally identifiable information used to identify specific Medicare beneficiaries and providers, including the beneficiary identifier (which can be a Social Security Number), provider identifier (providers can be identified using a Tax Identifier, which in some cases is a Social Security Number), date of birth, and geographic place of residence. The files available include paid institutional and non-institutional claims for Medicare Fee-for-Service beneficiaries, and enrollment data like the Master Beneficiary Summary File (MBSF). These datasets can be queried to extract information for a specific population of beneficiaries or providers. In addition, CMS creates files containing claims and enrollment data for samples of 5% of the Medicare beneficiary population, eliminating the need for researchers to request the full datasets (the size of which is overwhelming) and create their own samples. Because the RIFs contain personally-identifiable information, the use of such files is subjected to strict oversight by CMS, and access to these data is provided only when necessary. The data request for our sample research programming project will be from the RIFs.
2. **Limited Data Sets (LDS)**: Claims and enrollment files containing the same information provided in the RIFs except for certain confidential identifiers. Even with such exclusions, LDS files are considered personally identifiable datasets and are subject to strict privacy rules. LDS files can be requested by completing an LDS data request packet.
3. **Public Use Files (PUFs), or Non-Identifiable Files)**: These datasets generally contain summary-level, aggregated information on the Medicare program. Some research projects only require the aggregated results that are found in these non-identifiable files, eliminating the need for the researcher to acquire the data and personally aggregate the files. Note that requests for PUFs are made directly to CMS, though payment may still be required. Visit https://www.cms.gov/NonIdentifiableDataFiles/ for more details. In addition, note that CMS also provides synthetic PUFs for data entrepreneurs. See http://www.cms.gov/Research-Statistics-Data-and-Systems/Statistics-Trends-and-Reports/SynPUFs/DE_Syn_PUF.html for additional details.

In the case of our research programming project, we want the ability to report results by provider, necessitating the identification of providers. In addition, we will be looking at beneficiary-level information, necessitating data at the beneficiary-level. Of course, as we explained earlier, although we pretend to obtain our data from CMS's data distribution contractor, we actually use the synthetic PUFs for data entrepreneurs as the source data used for this book. Therefore, we are simply simulating the exercise of requesting, obtaining, and using personally identifiable data.

Our Project's Data Requirements and Request Specifications

Because it is relevant to understanding our work in future chapters, let's provide additional details on the personally identifiable data we are requesting, as well as the file we are using for our data request. In

Chapter 1, we discussed that the starting point for our example research programming project is a file provided by CMS that contains identifiers for the providers that participated in the example pilot program we are studying, along with identifiers for the beneficiaries associated with those providers. Associating providers and beneficiaries to assign responsibility for a beneficiary's care is called attribution. Constructing algorithms for provider attribution is outside of the scope of this text, so we will simply assume that this file of providers and associated beneficiaries provided by CMS can be used for provider attribution. From this point forward, we refer to this data set as a "finder file" (because we will send it to CMS's data distribution contractor upon request for their use in finding all of the claims for the beneficiaries listed in the file). We may also refer to this file as an "attribution file" (because in subsequent chapters we will use it to describe the providers associated with the beneficiaries in the file). Further, we assume this finder file will be provided to us as both a flat text file and a SAS data set (so we will not present code for reading the file into a SAS data set).

Our request to CMS must describe the need for identifying claims and enrollment information associated with the entities in our finder file. Therefore, we will ask CMS's data distribution contractor to extract claims and enrollment data for beneficiaries associated with the providers who participated in our example pilot program by providing the data distribution contractor with our finder file. Our data request will be for the following files for calendar year 2010: Part B carrier claims, DME claims (even though we will not use them in this book), inpatient claims, outpatient claims, SNF claims, hospice claims, home health claims, and MBSF enrollment data.[6] We will use the data received from CMS's data distribution contractor to develop algorithms to produce summaries of payment, utilization, and quality outcomes. In Chapter 10, we will use these analytic file summaries of beneficiary-level data to analyze the providers who participated in the pilot program (and we will use the aforementioned "attribution file" assumed to be provided by CMS to accomplish this task).

Contacting ResDAC and Completing Paperwork

It is important to keep in mind that ResDAC's mission is to assist researchers with CMS data. When you work on your own research programming project, you will need to contact ResDAC and obtain the latest instructions on how to execute a data request. Not only does ResDAC's expert staff answer questions about obtaining data, but they also answer questions about how to use the data. In other words, there are many resources for you to turn to in the data extraction process: from this book, to ResDAC, to fellow professional research systems specialists, to fellow researchers, to folks at CMS! The following high-level instructions for completing a data request to ResDAC are intended to introduce the topic of requesting data and are not a substitute for contacting ResDAC directly. ResDAC staff are the real experts, and the remainder of this chapter is no substitute for seeking their know-how.

Our first step is to visit the ResDAC and data distribution contractor's websites[7] to obtain the most up-to-date forms and information on data (like data dictionaries). We may wish to contact ResDAC to discuss our planned request, the research questions we need to answer, the data we believe we must request in order to answer those questions, and other pertinent information like the source of our funding. We may also wish to request an estimate of the cost to fulfill our data request at this time. (Depending on your source of funding, you may need to request a cost estimate prior to agreeing to undertake research just to ensure that your project is able to afford the data necessary to perform research. As with most tasks in project work, making contact and gathering information as early as possible typically pays large dividends down the road.)

Now that we have made initial contact and have an idea of what is contained in the request packet and the process we will need to work through, we can begin to fill out the paperwork. Walking through every step in the request is beyond the scope of this book (and not necessary given that ResDAC experts are available to provide guidance as you complete the request). Therefore, we will simply review the major steps in the process and highlight important information.

1. Complete a Data Use Agreement (CMS form CMS-R-0235, commonly referred to as a DUA, and contained in the request packet we downloaded from the ResDAC website). All requests for identifiable data must be accompanied by a DUA. A DUA is a contract between the users of the data (in the case of our research programming project, us!) and the owners of the data (in all cases, CMS). The contract specifies rules to keep the use of CMS's data in compliance with CMS's data use policies, including the Privacy Act of 1974, a federal law that covers the use of personally identifiable information. The DUA specifies the person responsible for the data (called the Data Custodian), as well as authorized users of the data. Use of the data is strictly limited to those individuals listed on the DUA (or who have a signed DUA signature addendum). The contract also specifies how to deal with the data at the close of the project (more on that in Chapter 10).
2. Complete the remainder of the data request packet. This collection of documents contains everything we need for our request as well as detailed instructions and checklists for completing the required forms. A precise and methodical approach to completing the request packet will minimize back-and-forth throughout the request process. For example, many of the documents in the request packet ask us to describe our research programming project and the identifiable files we are requesting. All such descriptive information should be consistent across all documents. The request packet includes the following documents:
    a. A request for the identifiable data on our organization's letterhead. A sample request letter is included in the request packet. The letter simply describes the purpose of our research programming project and how it relates to our need for the identifiable data files we are requesting.
    b. An Executive Summary form that asks us to describe our study, our plan for managing the data (e.g., how the data will be stored and destroyed and who will have access to it). ResDAC provides documents that guide the completion of the Executive Summary form.
    c. A Research Study Protocol Format form that asks us to describe the background, objectives, and methods of our study, as well as how the files included in our data request are necessary to meet these objectives.[8]
    d. Our completed Data Use Agreement form.
    e. Documentation that our request has been reviewed and approved by an Internal Review Board (IRB). This authorization is necessary for CMS to be able to release personally identifiable data.
    f. Documentation that our project has funding for the purchase of the data. The data can be expensive, so we must provide evidence that we have received funding internally from our own organization or from something like a grant or a contract with a federal agency.
    g. A Specification Worksheet that describes the user of the data, the data custodian, payment information, and information needed to perform the data extract. For example, we must

describe the operating system of the hardware we will use to process the data and the type of media on which we wish to receive the data.

h. A completed request for a CMS Cost Estimate. This document asks questions pertaining to the complexity of the data request.

3. Next, we must email a draft of the request packet to ResDAC for their review and commentary. It is important to allow plenty of time in your project schedule to receive ResDAC's comments on the draft materials, respond to those comments, and resend the modified draft request packet for approval.
4. Once ResDAC approves the draft request packet, ResDAC submits the materials to CMS for review and approval. The approval process contains many steps so it is important to allow plenty of time for the process to play out. For example, the request is assigned a control number and is subsequently presented to CMS's Privacy Board for review and approval. It is prudent to plan for a minimum of three or four weeks to pass between the submission of the final packet to CMS and getting word regarding the final approval of our request.
5. Once our request is reviewed and approved, we can make payment. It can take two weeks for the payment process to run its course.
6. Immediately upon submitting payment (we do not have to wait for confirmation), we can submit the instructions we would like to use for data extraction, including the aforementioned finder file on which our extraction will be based. However, processing will not begin until payment has been finalized by CMS.
7. Sit back, relax, and wait for your data to arrive! Go see a movie, get a massage, or take a vacation. Once CMS has finalized our payment and submitted our request to the data distribution contractor for processing, it can take at least 4-6 weeks to receive the data files we have been longing for!

Receiving, Decrypting, and Loading Medicare Administrative Claims and Enrollment Data

Receiving Our Medicare Administrative Data

Our patience has been rewarded; we have received the Medicare data we requested! We tear into the shipping package to reveal a mass storage device that contains our claims and enrollment data. We plug the storage device into our PC and find that it contains the following files:

- The requested data formatted as flat files with fixed columns compressed and encrypted into self-decrypting archive files (SDAs, recognizable by the file extension .dat)
- SAS code for loading the files into SAS datasets (recognizable by the file extension .sas)
- Summary information detailing the data transmission (recognizable by the file extension .fts, short for "file transfer summary")
- A host of documentation and data dictionaries including a 'readme' file listing the files that comprise the data transmission package, instructions for decrypting the data, spreadsheets detailing the variables included in each file, information on the diagnosis and procedure codes in the data, and data dictionaries

- Although not included in our package, remember that we have our finder file (or attribution file) for use in later chapters! We assume that this finder file is a SAS data set.

Decrypting Our Medicare Administrative Data

Our first hurdle is to uncompress and decrypt our data. We received our decryption key (a fancy way of saying "password") via email from CMS's data distribution contractor. One of the most common questions I receive relates to the inability to decrypt a file; this problem is usually rooted in accidentally attempting to decrypt the file to the media on which we received our data. Copying the files to our hard drive prior to attempting to decrypt the data avoids this problem. I recommend an additional step of copying all of the files included on the mass storage device into a directory and then making the directory read-only. This ensures we cannot accidentally modify our data, and we can work from this directory with confidence. Because the files are SDAs, we do not need to acquire any special decryption software; aside from the email containing our decryption key, everything we need for the decryption process is included in our shipment including an executable file (.exe) that runs the decryption process. We need simply to follow the instructions for decrypting the data files included in our data shipment.

Loading Our Medicare Administrative Data into SAS Data Sets

Once our files are copied and decrypted, we must load them into SAS datasets. As stated above, CMS's data distribution contractor makes the process as simple as possible by including SAS code and the instructions necessary to complete the task in our data shipment. Therefore, we will not spend time (nor will we provide code) discussing how to load our source files into SAS datasets, but it is worth the time to discuss a few details.

First, note that we received the following files:

- Base claim and revenue center files for all institutional claims (inpatient, outpatient, SNF, home health, and hospice). Recall that we discussed the difference between the base claim and the revenue center files in Chapter 3.[8]
- Base claim and claim line files for all non-institutional claims (DME and Carrier). Recall that we discussed the difference between the base claim and the line level files in Chapter 3.
- We also received the Master Beneficiary Summary File (MBSF) containing enrollment data. Recall that we discussed the MBSF in Chapter 3.

In addition, note that we received file transfer summary files (the aforementioned file with the extension .fts) and SAS program files that correspond in name to each of the claims and enrollment files. The transfer summary files contain information on the contents of each of our claims and enrollment files that confirm we loaded the data into SAS correctly. For example, the files contain information on the number of columns and rows, the length of the rows, and the size of the file. Once we load the flat files, the number of records in our SAS datasets should match the count of rows in the file transfer summary.

Finally, when loading the flat files into SAS datasets, we must include the proper SAS LIBNAMEs to ensure we are reading and writing to the proper locations (remembering that we assigned read-only rights to our source data directory). We must also be cognizant of how our file sizes will change as we move the data from compressed, encrypted files to flat files, and from flat files to SAS datasets. Planning ahead to ensure that we have the appropriate amount of storage allocated for our project is important. We listed the LIBNAMEs we will use for this project in Chapter 4. As a reminder, we will load the raw data received from CMS's data distribution contractor (as well as any other source data) into a library nicknamed SRC. We will transform this data and save it to a library nicknamed ETL.

After loading our claims and enrollment data into SAS data sets, we have the following files stored in our library with the alias SRC:

- Carrier base claim and line level files (SRC.CARRIER2010CLAIM and SRC.CARRIER2010LINE, respectively)
- DME base claim and line level files (SRC.DME2010CLAIM and SRC.DME2010LINE, respectively)
- Inpatient base claim and claim line files (SRC.IP2010CLAIM and SRC.IP2010LINE, respectively)
- Skilled Nursing Facility base claim and revenue center level files (SRC.SNF2010CLAIM and SRC.SNF2010LINE, respectively)
- Outpatient base claim and revenue center level files (SRC.OP2010CLAIM and SRC.OP2010LINE, respectively)
- Home health base claim and revenue center level files (SRC.HH2010CLAIM and SRC.HH2010LINE, respectively)
- Hospice base claim and revenue center level files (SRC.HS2010CLAIM and SRC.HS2010LINE, respectively)
- Master Beneficiary Summary File enrollment data (SRC.MBSF_AB_2010)
- Social Security Administration (SSA) state and county codes, and corresponding state and county names (more on this in Chapter 6)
- The finder file of provider identifiers and the beneficiaries attributed to those providers, received from CMS at the outset of the project and used in the extraction of our claims and enrollment data (SRC.PROV_BENE_FF)

The above data sets are available on the book's website (http://support.sas.com/publishing/authors/gillingham.html) and represent the starting point for our programming work. Note that the file names and variable names we use may differ from those provided by CMS's data distribution contractor.

Algorithms: Transforming Base Claim and Line Level Data Sets into Single Claim-Level Files

Let's start our programming work at the point after we have loaded the data we received from CMS's data distribution contractor into SAS data sets saved in the library with the alias SRC (again, we will not program the actual loading of the data because the data distribution contractor provides the programs for that task). At this point, we have base claim and revenue center level SAS data sets for each institutional claim type. In addition, we have base claim and line level SAS data sets for each non-institutional claim type. As mentioned above, we will combine the base claim and revenue center (or line) files into a single file per claim type, with one record for each claim and its associated claim lines.[9] Note that we will not provide code for executing this task for each claim type, nor will we provide code for using all variables in these data sets. Rather, we will provide a set of code for transforming data sets for Carrier claims, using a limited set of variables.[10] Using these examples, the reader can easily create code for performing this procedure on other claim types and additional variables.

Transforming Base Claim and Line-Level Carrier Data into a Claim-Level File

Let's combine our base claim and line level carrier SAS data sets into a single carrier SAS data set with one record for each claim and its associated claim lines. In Step 5.1, we sort the carrier line file (SRC.CARRIER2010LINE) by the variables **BENE_ID, CLM_ID**, and **CLM_LN**.

```
/* STEP 5.1: SORT CARRIER LINE FILE IN PREPARATION FOR TRANSFORMATION */
proc sort data=src.carrier2010line out=carrier2010line;
       by bene_id clm_id clm_ln;
run;
```

In Step 5.2, we transform the sorted carrier line file (CARRIER2010LINE) from one record per line (identified using the variable **CLM_LN)** to one record for each claim, with the claim line variables arrayed in that record. Because carrier claims can have up to 13 claim lines, we must create 13 new variables.[11] As you can see, we do the same for the other variables in our carrier line. These variables will be used in our work in later chapters.

```
/* STEP 5.2: TRANSFORM CARRIER LINE FILE */
data carrier2010line_wide(drop=i expnsdt1 expnsdt2 hcpcs_cd
line_icd_dgns_cd clm_ln linepmt prfnpi tax_num);
       format  expnsdt1_1-expnsdt1_13 mmddyy10.
                   expnsdt2_1-expnsdt2_13 mmddyy10.
                   line_icd_dgns_cd1- line_icd_dgns_cd13 $5.
                   hcpcs_cd1-hcpcs_cd13 $7.
                   linepmt1-linepmt13 10.2
                   prfnpi1-prfnpi13 $12.
                   tax_num1-tax_num13 $10.;
       set carrier2010line;
       by bene_id clm_id clm_ln;
       retain      expnsdt1_1-expnsdt1_13
                   expnsdt2_1-expnsdt2_13
                   line_icd_dgns_cd1- line_icd_dgns_cd13
                   hcpcs_cd1-hcpcs_cd13
                   linepmt1-linepmt13
                   prfnpi1-prfnpi13
                   tax_num1-tax_num13;

       array   xline_icd_dgns_cd(13) line_icd_dgns_cd1-
line_icd_dgns_cd13;
       array  xexpnsdt1_(13) expnsdt1_1-expnsdt1_13;
```

```
        array  xexpnsdt2_(13) expnsdt2_1-expnsdt2_13;
        array  xhcpcs_cd(13) hcpcs_cd1-hcpcs_cd13;
        array  xlinepmt(13) linepmt1-linepmt13;
        array  xprfnpi(13) prfnpi1-prfnpi13;
        array  xtax_num(13) tax_num1-tax_num13;

        if first.clm_id then do;
              do i=1 to 13;
                    xline_icd_dgns_cd(clm_ln)='';
                    xexpnsdt1_(clm_ln)=.;
                    xexpnsdt2_(clm_ln)=.;
                    xhcpcs_cd(clm_ln)='';
                    xlinepmt(clm_ln)=.;
                    xprfnpi(clm_ln)='';
                    xtax_num(clm_ln)='';
              end;
        end;

        xline_icd_dgns_cd(clm_ln)=line_icd_dgns_cd;
        xexpnsdt1_(clm_ln)=expnsdt1;
        xexpnsdt2_(clm_ln)=expnsdt2;
        xhcpcs_cd(clm_ln)=hcpcs_cd;
        xlinepmt(clm_ln)=linepmt;
        xprfnpi(clm_ln)=prfnpi;
        xtax_num(clm_ln)=tax_num;

        if last.clm_id then output;
run;
```

Figure 5.1 shows the observations for a specific **BENE_ID** in the CARRIER2010LINE data set. Note how the variables that describe a service, like diagnosis code (**LINE_ICD_DGNS_CD**) variable, are stored as individual records for different line numbers (**CLM_LN**) of the same claim (**CLM_ID**).

Figure 5.1: Sorted Carrier Line File

SORTED CARRIER LINE FILE

Obs	BENE_ID	CLM_ID	CLM_LN	LINE_ICD_DGNS_CD
4	00016F745862898F	887023385728164	1	56211
5	00016F745862898F	887023385728164	2	56211
6	00016F745862898F	887023385728164	3	33818

Figure 5.2 is a single observation taken from the CARRIER2010LINE_WIDE data set for the same beneficiary displayed above. You can see that the procedure code and diagnosis code variables formerly stored in individual rows are now arrayed in a single row.

Figure 5.2: Transformed Carrier Line File

TRANSFORMED CARRIER LINE FILE

Obs	BENE_ID	CLM_ID	hcpcs_cd1	hcpcs_cd2	hcpcs_cd3	line_icd_dgns_cd1	line_icd_dgns_cd2	line_icd_dgns_cd3
4	00016F745862898F	887023385728164	99223	4048F		56211	56211	33818

In Step 5.3, we sort the base claim SAS data set (SRC.CARRIER2010CLAIM) and the transformed claim line SAS data set (CARRIER2010LINE_WIDE) by **BENE_ID** and **CLM_ID** in preparation for merging the two files to create a carrier SAS data set that contains one record per base claim, with all associated claim lines. We save the output of the sort of the carrier base claim file, named CARRIER2010CLAIM, to our work directory.

```
/* STEP 5.3: SORT BASE CLAIM AND TRANSFORMED LINE FILES IN PREPARATION
FOR MERGE */
proc sort data=src.carrier2010claim out=carrier2010claim;
       by bene_id clm_id;
run;

proc sort data=carrier2010line_wide;
       by bene_id clm_id;
run;
```

Figure 5.3 is a single observation taken from the CARRIER2010CLAIM data set for the same beneficiary displayed for Step 5.3 above.

Figure 5.3: Sorted Carrier Base Claim File

SORTED CARRIER BASE CLAIM FILE

Obs	BENE_ID	CLM_ID	FROM_DT	THRU_DT
4	00016F745862898F	887023385728164	20100529	20100529

In Step 5.4, we merge the CARRIER2010CLAIM and CARRIER2010LINE_WIDE files by the unique beneficiary identifier (**BENE_ID**) and the unique claim identifier (**CLM_ID**), using an "if a and b" merge. We output a SAS data set called ETL.CARR_2010. This outputted carrier SAS data set, sorted by **BENE_ID**, will be used for identifying Part B carrier services throughout the remainder of this book. Note that we also output a file called CARR_NOMATCH that contains records that did not match by **BENE_ID** and **CLM_ID** for quality assurance purposes (mismatches indicate a potential issue with the data or your code).

```
/* STEP 5.4: MERGE BASE CLAIM AND TRANSFORMED LINE FILES */
data etl.carr_2010 carr_nomatch;
    merge carrier2010claim(in=a) carrier2010line_wide(in=b);
       by bene_id clm_id;
       if a and b then output etl.carr_2010;
       else output carr_nomatch;
run;
```

Figure 5.4 is a single record of the outputted ETL.CARR_2010 data set for the same beneficiary displayed above. You can see that the base claim and line files have been merged such that one record per claim exists, where variables like procedure code and diagnosis code formerly stored in individual rows in the line file are now arrayed in a single row, along with the base claim detail of the dates on the claim (**FROM_DT** and **THRU_DT**).

Figure 5.4: Merged Transformed Carrier Line and Base Claim Files

MERGED TRANSFORMED CARRIER LINE AND BASE CLAIM FILES

Obs	BENE_ID	CLM_ID	hcpcs_cd1	hcpcs_cd2	hcpcs_cd3	line_icd_dgns_cd1	line_icd_dgns_cd2	line_icd_dgns_cd3	FROM_DT	THRU_DT
4	00016F745862898F	887023385728164	99223	4048F		56211	56211	33818	20100529	20100529

Chapter Summary

In this chapter, we learned how to request, receive, load, and transform the Medicare administrative claims and enrollment data we will use to complete our research programming project. To this end, we discussed the following information:

- A data request can take a lot of time and cost a lot of money to fulfill, so it is important to plan as much as possible prior to initiating our data request.
- Depending on the source of your funding, there are several routes to procuring Medicare data for research purposes. We discussed CMS's Integrated Data Repository (IDR), Data Extract System (DESY), and Virtual Research Data Center (VRDC). In addition, we discussed the Research Data Assistance Center (ResDAC), and CMS's data distribution contractor.
- ResDAC and CMS's data distribution contractor are the research community's gateway to CMS data.

- Our instructional plan is to request our data from CMS's data distribution contractor and quickly convert the claims data we receive into separate files for each claim type that contain all of the information for a claim in one record.
- CMS offers a variety of datasets for research purposes which we grouped into three categories: Research Identifiable Files (RIFs), Limited Data Sets (LDS), and Public Use Files (PUFs). Consult with ResDAC to ensure that you are requesting the right kind of data for your research.
- Our data request utilized a finder file provided by CMS to query the following claims and enrollment data for calendar year 2010: Part B Carrier claims, DME claims, inpatient claims, outpatient claims, SNF claims, hospice claims, home health claims, and MBSF enrollment data.
- We discussed a high-level overview of the data acquisition process.
- The package of data we received from CMS's data distribution contractor contained our requested data (in encrypted files), the SAS code for loading the files into SAS datasets, summary information detailing the data transmission, and a host of documentation and data dictionaries.
- We decrypted our data files and loaded the files into SAS datasets. Then, we combined the administrative claims files, separated into files by claim type, header, and line information, into a set of files by claim type that contain all information on the paid claim in a single record. We retained access to the finder file (also called our attribution file) for use in subsequent chapters.

---

[1] As stated in Chapter 3, all discussions of acquisition of source data are for instructional purposes. The data we use in this book are fake, and subject to the disclaimer presented in Chapter 1. In other words, although we pretend to go through the process of requesting data from CMS, we are simply describing how the process would occur, and we did not actually procure the data used in this book through that process.

[2] See, for example, ResDAC's Data Request for Student Research webpage, available at http://www.resdac.org/data-request-student-research.

[3] The list of data sources is not exhaustive, and new sources of data do become available.

[4] For instructional purposes, as part of our analytic work we will transpose some of the files we create in this chapter to create a file containing one record per line.

[5] All this said, it is important to note that there are advantages to maintaining separate claim header and claim line data sets, and using PROC SQL to join the data sets when necessary. In your work, and consistent with our discussion about planning, you should consider the most efficient way to arrange your data and write your SAS code.

[6] The reader may wish to request a prior or subsequent year of data to study some measures like hospital readmissions, or to make comparisons to prior years of performance outcomes (i.e., compare a provider's performance in 2010 to her performance in 2009). For our simple examples, this will not be necessary.

[7] See www.resdac.org and www.ccwdata.org, respectively.

[8] For a federally funded study, the methods section should be included rather than writing up a new protocol per the packet sample.

[9] The institutional files can also include other reference code files for things like condition and span codes. We will not use these files, so we will not discuss them further.

[10] Although it is important to be aware of the fact that inpatient claims can spill over into more than one claim record, it does not happen often, so we will not write code to handle this situation.

[11] Our code and explanation draws on a helpful SAS Learning Module provided by UCLA's Institute for Digital Research and Education, available at http://www.ats.ucla.edu/stat/sas/modules/longtowide_data.htm.

Chapter 6: Working with Enrollment Data

Introduction and Goals

In Chapter 5, we acquired the Medicare enrollment file for the 2010 calendar year and loaded the file into a SAS data set. In this chapter, our goal is to utilize this Master Beneficiary Summary File (MBSF) data set in our research programming project. As discussed in earlier chapters, the online companion to this book is at http://support.sas.com/publishing/authors/gillingham.html. Here, you will find information on creating dummy source data, the code in this and subsequent chapters, as well as answers to the exercises in this book. I expect you to visit the book's website, create your own dummy source data, and run the code yourself.

Recall that we have been asked to evaluate the 2010 outcomes of a pilot program designed to incentivize providers to reduce costs and improve quality. In order to do so, we will compute simple measures of payment, utilization, and quality outcomes for those providers that interacted with the beneficiaries in our sample population. The pilot program operated for the full calendar year of 2010. Although a population of providers (and associated beneficiaries) participating in the pilot program was chosen at the outset of the demonstration, we have been asked to perform our analyses only on those beneficiaries who were continuously enrolled in Medicare Fee-for-Service (FFS) throughout all twelve months of 2010. Therefore, we must write and execute SAS code that delimits our beneficiary population to those beneficiaries continuously enrolled in Medicare FFS throughout all twelve months of calendar year 2010.[1] In subsequent chapters we will also use this file to delimit our calendar year 2010 claims data. We will then develop algorithms that will query the claims data to produce summaries of payment, utilization, and quality. Given this plan, we will use the MBSF data to do the following:

- Identify beneficiaries continuously enrolled in Medicare FFS throughout all twelve months of calendar year 2010, and use this information to delimit our 2010 MBSF data set to continuously enrolled beneficiaries.
- Create or retain data elements for displaying results by the following beneficiary characteristics:
  - Medicare coverage characteristics in 2010
  - Sex
  - Age groups (less than 65, 65-74, 75-84, 85-94, and 95 and older)
  - Beneficiary geographic information, defined by Social Security Administration (SSA) state and county codes, along with the corresponding state and county names
- Create a file of beneficiary enrollment and demographic information distilled from the MBSF data for 2010 that we will use for the remainder of our programming for our sample research project. This file will be at the beneficiary level, with one record per beneficiary.

Review and Approach

Basics of Medicare Enrollment Data

In Chapter 2 through Chapter 5, we discussed the following characteristics of Medicare enrollment, the Master Beneficiary Summary File, and programming with Medicare administrative data:

- The majority of Medicare beneficiaries are eligible for Medicare insurance because they are aged 65 and over. However, under certain circumstances, Medicare also insures beneficiaries who are permanently disabled or have ESRD or ALS.
- Beneficiaries can enroll in Medicare Parts A and B or Medicare Part C. Medicare Parts A and B are FFS programs, while Medicare Part C is a managed care program (Medicare Advantage).
- Medicare Part C permits Medicare beneficiaries to enroll in a managed care organization instead of participating in traditional Medicare FFS. Medicare Part A helps pay for care that is provided in an institutional setting, like inpatient hospitals. Medicare Part B helps pay for care that is

provided in a non-institutional setting, like a physician's office, as well as institutional outpatient services.

- Generally, the claims of beneficiaries enrolled in Part C do not appear in the administrative claims data, but some HMOs do report claims and these claims can appear in the administrative data (although in such a very small number that we cannot study these beneficiaries because we will not have full utilization information).
- The Medicare MBSF data contains information on Medicare beneficiaries, including enrollment and demographic characteristics like reason for entitlement, date of birth, age, sex, state, county, and zip code.
- MBSF data are available by request. Our example enrollment data for 2010 arrived in flat file format and we loaded it into a SAS dataset named SRC.MBSF_AB_2010.
- The MBSF data dictionary defines the variables in the file. The data dictionary is available from CMS's data distribution contractor and is integral to the effective use of the MBSF.
- It is prudent to plan effectively prior to beginning your programming project. Part of this planning effort involves creating written specifications that will guide the creation of your SAS algorithms. In addition, it is equally important to execute quality assurance and quality control steps throughout your programming process. These QA/QC procedures include reviewing the written specifications and debugging your SAS code through viewing output and test cases, as well as benchmarking output. Finally, because Medicare administrative data files can be quite large, it is important to keep efficient programming techniques in mind when coding.

Our Programming Plan

We will use the 2010 MBSF enrollment data to establish our population of full-year FFS beneficiaries (i.e., those beneficiaries enrolled in Medicare FFS throughout all twelve months of 2010). We will also utilize the 2010 MBSF for all enrollment information needed for our research programming effort. Therefore, our plan is to first query the 2010 MBSF and establish our population of full-year FFS beneficiaries. Next, we will turn our attention to creating descriptive variables. With this in mind, we can begin our programming with MBSF data by identifying beneficiaries in our sample population who were continuously enrolled in Medicare FFS throughout all twelve months of calendar year 2010.

Algorithms: Identifying Continuously Enrolled FFS Beneficiaries

Why Define Continuously Enrolled FFS Beneficiaries?

Identifying beneficiaries who have been continuously enrolled in Medicare FFS (be it for a calendar year or a period prior to or following a specific medical event, like a hospitalization) is a common task in research programming. Because Medicare administrative claims data include all FFS claims submitted on behalf of a beneficiary (and most likely will not include claims for managed care enrollees), focusing on beneficiaries who are enrolled in Medicare FFS helps to ensure that we have a complete picture of each beneficiary's medical history for that year. In our case, we are seeking to determine the impact of a program in the year 2010 on payment, utilization, and quality. In order to do so comprehensively, we will limit the population we are studying to those beneficiaries who have been continuously enrolled in

Medicare FFS for all twelve months of the 2010 calendar year to help ensure that we have an accurate picture of their claims data during that time.

How to Specify the Programming for Continuous Enrollment in Medicare FFS

Depending on your project needs, "continuous enrollment" can be defined in many different ways; it is just a business rule. We will define continuous enrollment as follows: A beneficiary is considered continuously enrolled in Medicare FFS throughout 2010 if they were alive for the entire year, did not have managed care coverage at any point during the year, and were enrolled in Medicare Parts A and B all year.[2] Accomplishing this task requires understanding and using the following variables in the Master Beneficiary Summary File, which will be explained in detail in the remainder of this chapter:

- Medicare Hospital Insurance (Part A) and Supplementary Medical Insurance (Part B) coverage variables
- HMO coverage variable
- Date of death variable

Medicare Part A and Part B Enrollment Variables

The Medicare Part A and Part B Enrollment variables (**BENE_HI_CVRAGE_TOT_MONS** and **BENE_SMI_CVRAGE_TOT_MONS**) describe the number of months that a beneficiary is enrolled in Part A and Part B. These variables are sometimes referred to by the shortened variable names **A_MO_CNT** and **B_MO_CNT**. It is important to determine enrollment because services received when a beneficiary is not enrolled in Medicare Part A or Part B for a period of time are not contained in the claims files. Therefore, a distorted picture of, say, utilization may result if gaps in coverage are allowed.

Checking the data dictionary for the MBSF[3], we want to keep only those beneficiaries with values of 12 for each variable, denoting that the beneficiary was enrolled in Part A and Part B for all 12 months of the year. We do not want to keep beneficiaries with a value of less than 12 for either the **_HI_CVRAGE_TOT_MONS** variable or the **_SMI_CVRAGE_TOT_MONS** variable because we stated above that we want to retain only those beneficiaries who have been enrolled in both Medicare Part A and Part B throughout 2010.

HMO Coverage Variable

The HMO coverage variable (**BENE_HMO_CVRAGE_TOT_MONS**) describes the number of months that a beneficiary was enrolled in a managed care plan (also known as a Health Maintenance Organization, or HMO) during the year. As stated above, Medicare Part C permits Medicare beneficiaries to enroll in a private managed care insurance plan instead of participating in traditional Medicare FFS. Generally, the claims of beneficiaries enrolled in these managed care plans do not appear in the administrative claims data, so we seek to exclude these beneficiaries from our study population. Another quick check of the MBSF data dictionary shows we want to keep only those beneficiaries with a value of 0 for the **BENE_HMO_CVRAGE_TOT_MONS** variable.

All other values indicate that the beneficiary was enrolled in managed care and what entity will process the beneficiary's claims.

Date of Death Variable

The date of death variable (**DEATH_DT**) describes the date that a beneficiary died. We want to keep only those beneficiaries who were alive in 2010. Although we will not use it, the date of death variable can also be confirmed by using the Valid Date of Death Switch variable (**V_DOD_SW**); this variable indicates that the beneficiary's date of death has been confirmed by the Social Security Administration as accurate.[4] In this case, we will choose to delimit our population to records where the **DEATH_DT** variable is equal to missing, indicating that no date of death is available.

How to Program in SAS to Define Continuous Enrollment in Medicare FFS

Now that we have specified how to define continuous enrollment, we can build the code. The great thing about SAS is that there are many, many different ways to accomplish a single task, and there is really no right or wrong method (as long as the different approaches provide the same, correct answer!). That said, we are often concerned with efficiency when using Medicare data because the data sets can be quite large. It is important to get into the habit of reading and writing as few times as possible. Therefore, we will attempt to perform the tasks listed above in as few DATA steps as possible, while creating code that is (hopefully) useful from an instructional perspective. In addition, we will take a first step in quality assuring our code by examining our output and attempting to replicate our results.

I present the following method for identifying and retaining beneficiaries who have been continuously enrolled in Medicare FFS. First, in Step 6.1, we perform a simple data step to flag those beneficiaries continuously enrolled in Medicare Parts A and B. Specifically, we create a variable called **CONTENRL_AB_2010** that is set equal to 'AB' if the values of **BENE_HI_CVRAGE_TOT_MONS** and **BENE_SMI_CVRAGE_TOT_MONS** are both equal to 12, denoting that the beneficiary was enrolled in Medicare Parts A and B for all twelve months of the year. In the same fashion, we create a variable called **CONTENRL_HMO_2010** that is set equal to 'NOHMO' if the value of **BENE_HMO_CVRAGE_TOT_MONS** is equal to 0, denoting that the beneficiary was not enrolled in an HMO at any time during the year. Finally, we create a variable called **DEATH_2010** that is set equal to 0 if the value of **DEATH_DT** is null, denoting that the beneficiary was alive during all twelve months of year.

```
/* STEP 6.1: BUILD CONTINUOUS ENROLLMENT INFORMATION IN 2010 MBSF FILE
*/
data enr.contenr_2010;
    set src.mbsf_ab_2010;
       length contenrl_ab_2010 contenrl_hmo_2010 $5.;
       /* FLAG BENEFICIARIES WITH PARTS A AND B OR HMO COVERAGE */
    if bene_hi_cvrage_tot_mons=12 and bene_smi_cvrage_tot_mons=12 then
contenrl_ab_2010='ab'; else contenrl_ab_2010='noab';
    if bene_hmo_cvrage_tot_mons=12 then contenrl_hmo_2010='hmo'; else
contenrl_hmo_2010='nohmo';
```

```
        /* FLAG BENEFICIARIES THAT DIED IN 2010 */
        if death_dt ne . then death_2010=1; else death_2010=0;
run;
```

In Step 6.2, we output the following frequency distribution of the enrollment flags we created in Step 6.1, the results of which are illustrated by Output 6.1:

```
/* STEP 6.2: FREQUENCY OF CONTINUOUS ENROLLMENT VARIABLES */
ods html file="C:\Users\mgillingham\Desktop\SAS
Book\FINAL_DATA\ODS_OUTPUT\Gillingham_fig6_2_ENRL.html"
image_dpi=300 style=GrayscalePrinter;
ods graphics on / imagefmt=png;
title "VARIABLES USED TO DETERMINE CONTINUOUS ENROLLMENT IN 2010 DATA";
proc freq data=enr.contenr_2010;
    tables contenrl_ab_2010 contenrl_hmo_2010 death_2010 / missing;
run;
ods html close;
```

Output 6.1: Continuous Enrollment Variables

VARIABLES USED TO DETERMINE CONTINUOUS ENROLLMENT IN 2010 DATA

The FREQ Procedure

contenrl_ab_2010	Frequency	Percent	Cumulative Frequency	Cumulative Percent
ab	97319	86.31	97319	86.31
noab	15435	13.69	112754	100.00

contenrl_hmo_2010	Frequency	Percent	Cumulative Frequency	Cumulative Percent
hmo	30832	27.34	30832	27.34
nohmo	81922	72.66	112754	100.00

death_2010	Frequency	Percent	Cumulative Frequency	Cumulative Percent
0	110891	98.35	110891	98.35
1	1863	1.65	112754	100.00

Finally, in Step 6.3 we create our file of continuously enrolled beneficiaries by delimiting the **ENR.CONTENR_2010** file created in Step 6.1 by the enrollment flags we defined in that same step. A beneficiary is defined as continuously enrolled in Medicare Parts A and B in calendar year 2010 if the value of **CONTENRL_AB_2010** is equal to 'AB,' the value of **CONTENRL_HMO_2010** is equal to 'NOHMO,' and the value of **DEATH_2010** is not equal to 1. We keep only these records.

```
/* STEP 6.3: CREATE A 2010 ENROLLMENT FILE OF ONLY CONTINUOUSLY ENROLLED
BENEFICIARIES */
data enr.contenr_2010_fnl;
    set enr.contenr_2010;
       if contenrl_ab_2010='ab' and contenrl_hmo_2010='nohmo' and
death_2010 ne 1;
run;
```

Algorithms: Create or Retain Data Elements for Displaying Results by Certain Characteristics

In addition to delimiting our population to beneficiaries who were continuously enrolled in Medicare FFS throughout 2010, we also must create flags for displaying our results in later chapters by the following beneficiary characteristics:

- Medicare coverage characteristics in 2010
- Sex
- Race
- Age groups (less than 65, 65-74, 75-84, 85-94, and 95 and older)
- Beneficiary geographic information, defined by assigning an SSA state and county code, and corresponding state and county names, to each beneficiary

While we are working with the MBSF data in this chapter, we will create or retain these descriptive data elements. Although we cannot utilize these data elements to segment our payment and utilization statistics until later chapters, it is helpful to create them in this chapter while we are working with the enrollment data. We can put the variables to immediate use by using them to study the composition of our population. Accomplishing this now does not really influence the efficiency of our code (we can just set the output aside until needed), and it comes with the benefit of knowing more about our study population and finalizing the MBSF information we will need to complete the remainder of the programming for our sample project. This will be well worth the effort!

Coverage Characteristics, Month of Death, Sex, and Race

Our input dataset is the output of the continuous enrollment exercise, ENR.CONTENR_2010_FNL. We do not need to create separate flags to define enrollment characteristics in the 2010 data because we can simply use the variables created in Step 6.2. In addition, there is no need to perform any programming to create separate variables that contain information on sex or race; the **SEX** and **RACE** variables provided in the MBSF files serve our purposes. In Step 6.4, we simply explore the information contained in the

SEX and **RACE** variables using frequency distributions. Note that we use PROC FORMAT so the definitions of the values of **SEX** and **RACE** are displayed in our output. We use SAS' Output Delivery System (ODS) to create our output.

```
/* STEP 6.4: INITIAL INVESTIGATION OF SEX AND RACE IN THE 2010 DATA */
proc format;
    value $sex_cats_fmt
              '0'='UNKNOWN'
        '1'='MALE'
        '2'='FEMALE';
run;

ods html file="C:\Users\mgillingham\Desktop\SAS
Book\FINAL_DATA\ODS_OUTPUT\Gillingham_fig6_4_SEX.html"
image_dpi=300 style=GrayscalePrinter;
ods graphics on / imagefmt=png;
title "FREQUENCY OF SEX IN 2010 DATA";
proc freq data=enr.contenr_2010_fnl;
    tables sex / missing;
       format sex $sex_cats_fmt.;
run;
ods html close;

proc format;
    value $race_cats_fmt
        '0'='UNKNOWN'
        '1'='WHITE'
        '2'='BLACK'
        '3'='OTHER'
        '4'='ASIAN'
              '5'='HISPANIC'
              '6'='NORTH AMERICAN NATIVE';
run;

ods html file="C:\Users\mgillingham\Desktop\SAS
Book\FINAL_DATA\ODS_OUTPUT\Gillingham_fig6_4_RACE.html"
image_dpi=300 style=GrayscalePrinter;
ods graphics on / imagefmt=png;
title "FREQUENCY OF RACE IN 2010 DATA";
proc freq data=enr.contenr_2010_fnl;
    tables race / missing;
       format race $race_cats_fmt.;
run;
ods html close;
```

Output 6.2 and Output 6.3 show the results of Step 6.4. You can see that there are more females than males in our population, and that the majority of our beneficiaries are white.

Output 6.2: Percentage of Sex in 2010 Data

FREQUENCY OF SEX IN 2010 DATA

The FREQ Procedure

Sex				
SEX	Frequency	Percent	Cumulative Frequency	Cumulative Percent
MALE	29627	44.09	29627	44.09
FEMALE	37571	55.91	67198	100.00

Output 6.3: Percentage of Race in 2010 Data

FREQUENCY OF RACE IN 2010 DATA

The FREQ Procedure

Beneficiary Race Code				
RACE	Frequency	Percent	Cumulative Frequency	Cumulative Percent
WHITE	56409	83.94	56409	83.94
BLACK	6652	9.90	63061	93.84
OTHER	2663	3.96	65724	97.81
HISPANIC	1474	2.19	67198	100.00

Age Groups

Next, let's define and group the age category to which a beneficiary belongs. Suppose for illustrative purposes we wish to calculate each beneficiary's age as of the beginning of the reference year (in our case, January 1, 2010). In Step 6.5, we use our ENR.CONTENR_2010_FNL data set to calculate this variable and call it **STUDY_AGE**. Suppose further that we wish to create a variable that describes a beneficiary's inclusion in one of four age groups: 65 and younger, 65-74, 75-84, 85-94, and 95 and older.[5] Also in Step 6.5, we create a variable called **AGE_CATS** that groups beneficiaries by their value of **STUDY_AGE**. Note that we use PROC FORMAT to create a format called AGE_CATS_FMT that we apply to the display of **AGE_CATS** below.

```
/* STEP 6.5: CREATE VARIABLE NAMED STUDY_AGE THAT CONTAINS AGE AS OF
01.01.2010 */
/* STEP 6.5 (CONT): CREATE VARIABLE AGE_CATS THAT GROUPS STUDY_AGE INTO
AGE CATEGORIES */
proc format;
    value age_cats_fmt
        0='AGE LESS THAN 65'
        1='AGE BETWEEN 65 AND 74, INCLUSIVE'
        2='AGE BETWEEN 75 AND 84, INCLUSIVE'
        3='AGE BETWEEN 85 AND 94, INCLUSIVE'
        4='AGE GREATER THAN OR EQUAL TO 95';
run;

data enr.contenr_2010_fnl;
    set enr.contenr_2010_fnl;
    format age_cats age_cats_fmt.;
    study_age=floor((intck('month', bene_dob, '01jan2010'd) -
(day('01jan2010'd) < day(bene_dob))) / 12);
    select;
        when (study_age<65)        age_cats=0;
        when (65<=study_age<=74) age_cats=1;
        when (75<=study_age<=84) age_cats=2;
        when (85<=study_age<=94) age_cats=3;
        when (study_age>=95)       age_cats=4;
       end;
    label age_cats='Beneficiary age category at beginning of reference
year (January 1, 2010)';
run;
```

Finally, in Step 6.6, Output 6.4 shows the results of Step 6.5 (still using the file ENR.CONTENR_2010_FNL) using a cross-tabulation of **STUDY_AGE** (the variable we calculated as of January 1, 2010), and **AGE_CATS** (the groupings of **STUDY_AGE**). The output of this frequency can be used to check that our calculations of **STUDY_AGE** and **AGE_CATS** were performed correctly. Note that we delimited our input data to retain only those beneficiaries aged 65 through 70, inclusive for display purposes below. You can remove the 'where' clause and perform a frequency distribution for all values of **STUDY_AGE** at your discretion.

```
/* STEP 6.6: DISPLAY AGE GROUP CHARACTERISTICS IN 2010 ENROLLMENT DATA
*/
ods html file="C:\Users\mgillingham\Desktop\SAS
Book\FINAL_DATA\ODS_OUTPUT\Gillingham_fig6_6_AGE.html"
image_dpi=300 style=GrayscalePrinter;
ods graphics on / imagefmt=png;
title "CROSS TAB OF STUDY_AGE AND AGE_CATS IN 2010 DATA";
proc freq data=enr.contenr_2010_fnl(where=(65<=study_age<=70));
    tables study_age * age_cats / list missing;
    format age_cats age_cats_fmt.;
run;
ods html close;
```

Here is the output of frequency distribution executed in Step 6.6. You can see the computation of **AGE_CATS** is accurate.

Output 6.4: Age Groups

CROSS TAB OF STUDY_AGE AND AGE_CATS IN 2010 DATA

The FREQ Procedure

study_age	age_cats	Frequency	Percent	Cumulative Frequency	Cumulative Percent
65	AGE BETWEEN 65 AND 74, INCLUSIVE	514	3.27	514	3.27
66	AGE BETWEEN 65 AND 74, INCLUSIVE	2852	18.13	3366	21.40
67	AGE BETWEEN 65 AND 74, INCLUSIVE	3051	19.39	6417	40.79
68	AGE BETWEEN 65 AND 74, INCLUSIVE	3148	20.01	9565	60.80
69	AGE BETWEEN 65 AND 74, INCLUSIVE	3072	19.53	12637	80.33
70	AGE BETWEEN 65 AND 74, INCLUSIVE	3094	19.67	15731	100.00

Geographic Characteristics

What are some effective ways to display healthcare data by geographic characteristics? Certainly, grouping by state may be effective, but may also yield results that are defined too broadly. On the other hand, viewing by zip code may be too fine a level of granularity to pull out any meaningful results. In some cases, it may be informative to designate Hospital Referral Regions (HRRs) or Hospital Service Areas (HSAs) [6], or to assign a Metropolitan Statistical Area (MSA).

In our example research programming project, we will add the SSA state and county name to our enrollment data set. SSA state and county codes already exist on the MBSF data we received and loaded in Chapter 5.[7] A publicly available SSA code file contains the state and county names that correspond to the SSA state and county codes.[8] In order to add the state and county names to the MBSF data, we must load the SSA code file into SAS and work to merge the county names onto our MBSF data. In other words, our task is to merge state and county names onto our data set that contains enrollment information for those beneficiaries who have been continuously enrolled in Medicare FFS throughout all twelve months of 2010 (called ENR.CONTENR_2010_FNL). We do this by merging our enrollment data with the SSA code file by SSA state and county codes.

In Step 6.7, we begin by loading the SSA code file (called MSABEA.TXT) into a SAS data set. To this end, we create a SAS data set called SRC.MSABEA_SSA that contains the SSA state and county codes (concatenated into a single variable called **SSA**), the corresponding county name (**COUNTY**), and the corresponding state name (**STATE**).

```
/* STEP 6.7: LOAD SSA STATE AND COUNTY CODE INFORMATION */
data src.msabea_ssa;
      infile "C:\Users\mgillingham\Desktop\SAS
Book\FINAL_DATA\source_data\MSABEA03.TXT" missover;
      input
            county $  1-25
            state  $ 26-27
            ssa    $ 30-34;
run;
```

In Step 6.8, we prepare to merge this file with our MBSF data by sorting the data set by the **SSA** variable.[9]

```
/* STEP 6.8: SORT SSA STATE AND COUNTY CODES FILE TO REMOVE DUPLICATE
RECORD FOR DADE OR MIAMI DADE */
proc sort data=src.msabea_ssa nodupkey;
      by ssa;
run;
```

In Step 6.9, we prepare our enrollment file (SRC.CONTENR_2010_FNL) to receive the SSA state and county name information loaded into the SRC.MSABEA_SSA data set. Prior to merging with the SRC.MSABEA_SSA data set, we must create a variable on the SRC.CONTENR_2010_FNL data set that is equivalent to the **SSA** variable on the SRC.MSABEA_SSA data set. More specifically, our ENR.CONTENR_2010_FNL data set contains the information stored in the SSA variable in two variables (**STATE_CD** and **CNTY_CD**). Therefore, in order to merge our enrollment data with the SSA code file, we concatenate these two separate variables, creating an equivalent **SSA** variable (also called **SSA**) on the ENR.CONTENR_2010_FNL data set.

```
/* STEP 6.9: CREATE SSA VARIABLE ON ENROLLMENT DATA */
data enr.contenr_2010_fnl;
      set enr.contenr_2010_fnl;
      ssa=state_cd||cnty_cd;
run;
```

In Step 6.10, we proceed to merge the county and state names onto our enrollment data. First, we sort the ENR.CONTENR_2010_FNL data set by SSA, and then we perform a simple merge of the ENR.CONTENR_2010_FNL and the SRC.MSABEA_SSA data sets by **SSA**, keeping all of the data in our enrollment data set. In this way, we have assigned a state and county name to each beneficiary in our file of beneficiaries who were continuously enrolled in Medicare FFS throughout all twelve months of 2010.

```
/* STEP 6.10: SORT CONTINUOUS ENROLLMENT DATA AND MERGE WITH MSABEA FILE
*/
proc sort data=enr.contenr_2010_fnl; by ssa; run;

data enr.contenr_2010_fnl;
       merge enr.contenr_2010_fnl(in=a) src.msabea_ssa(in=b);
       by ssa;
       if a;
run;
```

Finally, in Step 6.11, we perform a simple print of the newly added state and county name variables in the ENR.CONTENR_2010_FNL data set (displaying just the first 10 records).

```
/* STEP 6.11: DISPLAY SSA STATE AND COUNTY NAMES IN 2010 ENROLLMENT DATA
*/
ods html file="C:\Users\mgillingham\Desktop\SAS
Book\FINAL_DATA\ODS_OUTPUT\Gillingham_fig6_11.html"
image_dpi=300 style=GrayscalePrinter;
ods graphics on / imagefmt=png;
title "SSA STATE AND COUNTY NAMES IN 2010 DATA";
proc print data=enr.contenr_2010_fnl(obs=10);
       var bene_id ssa state county;
run;
ods html close;
```

We use ODS to display the first 10 records of the output of Step 6.11, shown below in Output 6.5.

Output 6.5: SSA State and County Names

SSA STATE AND COUNTY NAMES IN 2010 DATA

Obs	BENE_ID	ssa	state	county
1	331B522FC848AC4A	01000	AL	AUTAUGA
2	3BA6CAAAE365F744	01000	AL	AUTAUGA
3	401FA84E59150F7B	01000	AL	AUTAUGA
4	52DA53C348BC581E	01000	AL	AUTAUGA
5	57ED3C3E76249A0D	01000	AL	AUTAUGA
6	5F2E2A7BA489D093	01000	AL	AUTAUGA
7	649C6458650337A2	01000	AL	AUTAUGA
8	9BED4955980743EC	01000	AL	AUTAUGA
9	B5C863C3DB3D0D54	01000	AL	AUTAUGA
10	FE68EC62AA8B3519	01000	AL	AUTAUGA

Algorithms: Create Final Enrollment Data for Remainder of Programming

The last step in our programming with MBSF data is to create a final file that we will carry forward and use throughout the remainder of our programming. In Step 6.12, we merely need to carry the file ENR.CONTENR_2010_FNL through the remainder of our processing. Since we will be doing much of our work at the beneficiary level, the last programming step in this chapter is to sort the ENR.CONTENR_2010_FNL data set by the beneficiary identifier **BENE_ID**.

```
/* STEP 6.12: CREATE FINAL ENROLLMENT FILE */
proc sort data=enr.contenr_2010_fnl;
       by bene_id;
run;
```

Chapter Summary

In this chapter we used the 2010 MBSF data sets to begin our research programming project. We:

- Specified our continuous enrollment criteria and wrote code to identify beneficiaries who were continuously enrolled in Medicare FFS throughout calendar year 2010. We used this information to delimit our 2010 MBSF data.
- Created and retained data elements for displaying results by Medicare coverage characteristics, sex, race, age groupings, and state and county name.
- Learned about commonly used enrollment data elements like the date of death, Parts A and B coverage, HMO enrollment, and beneficiary date of birth.
- Wrote an algorithm to calculate beneficiary age and create beneficiary age categories.
- Learned about SSA state and county codes, and created an algorithm to merge state and county name information onto our enrollment data.
- Created a final analytic file of enrollment information that we will use for the remainder of our programming for our sample research project.

Exercises

1. Alter the continuous enrollment specifications and code to flag beneficiaries who were continuously enrolled for any six consecutive months in 2010.
2. Using the 2010 MBSF data, write code to retain beneficiaries who reside in the following states: Pennsylvania, District of Columbia, Maryland, Michigan, Ohio, and Virginia. What variable describes the beneficiary's state of residence? Be sure to include a frequency to present your results.
3. In this chapter, we wrote our code in piecemeal fashion for educational purposes. We will use the same steps and illustrative coding process in later chapters. For your own projects, you may want to combine some of the code into fewer steps for efficiency reasons. Can you rewrite the code in this chapter so it is more efficient? How would you measure efficiency?

---

[1] In reality, we would have requested our data from CMS with the restriction of asking CMS's data distribution contractor to provide only claims for those beneficiaries who were enrolled in Medicare FFS throughout all twelve months of 2010. However, for instructional purposes, we will assume that we must perform this restriction ourselves.

[2] Often, researchers also exclude Medicare Secondary Payer and ESRD beneficiaries.

[3] Data dictionaries are available on the Chronic Conditions Data Warehouse website at https://www.ccwdata.org/web/guest/data-dictionaries.

[4] If a beneficiary's day of death cannot be confirmed, then it is assigned as the last day of the month of death. For more information, see ResDAC's article on this issue available on ResDAC's website at http://www.resdac.org/resconnect/articles/117.

[5] We assume that there are no missing values of the **STUDY_AGE** variable. If missing values did exist, they would end up in the "65 and younger" category.

[6] For more information on HRRs and HSAs, see The Dartmouth Atlas of Healthcare website at http://www.dartmouthatlas.org/data/region/.

[7] It is important to note the existence of multiple state and county coding systems. While the MBSF data uses the SSA state and county coding system, other sources may use the Federal Information Processing Standard (FIPs) coding system.

[8] The SSA code file used in this book (called MSABEA.TXT) is available on CMS's MSABEA file webpage available at http://www.cms.gov/Medicare/Medicare-Fee-for-Service-Payment/AcuteInpatientPPS/Acute-Inpatient-Files-for-Download-Items/CMS022639.html.

[9] Note that we perform a nodupkey sort to remove a record that is duplicative for our purposes.

Chapter 7: Measuring Utilization of Services

Introduction and Goals

Service utilization is an important component of our work because measuring the use of healthcare services is necessary to determine things like patterns of care, access to care, resource use, and quality of care. Our research programming project requires us to calculate utilization measurements to evaluate the outcomes of a program that is designed to incentivize providers to reduce costs and improve quality. Measuring the utilization characteristics of our provider population helps us to understand the lay of the land so that we can better comment on why an incentive payment may or may not have worked. For example, perhaps an incentive payment did not work because of geographic differences in how medicine is practiced. If that were the case, one outcome of our evaluation would be to recommend investigating medical education and methods of practice in order to reduce costs.

Therefore, we will execute some relatively straightforward utilization algorithms in this chapter, both for the sake of performing the calculations planned for this chapter as well as setting up measurements calculated in later chapters, such as payment and quality outcomes for certain chronic conditions. To these ends, in this chapter, we will code algorithms to do the following:

- Delimit our administrative claims data to include only those claims for the population of interest, those continuously enrolled beneficiaries formed in Chapter 6.
- Measure evaluation and management (E&M) visits in a physician office setting.
- Measure inpatient hospital utilization.
- Measure the professional component of emergency department (ED) services.
- Measure the utilization of ambulance services using the Part B carrier data.
- Measure outpatient visit utilization.
- Measure utilization of skilled nursing facility (SNF), home health, and hospice care.

As discussed in earlier chapters, the online companion to this book is at http://support.sas.com/publishing/authors/gillingham.html. Here, you will find information on creating dummy source data, the code in this and subsequent chapters, as well as answers to the exercises in this book. I expect you to visit the book's website, create your own dummy source data, and run the code yourself.

Before we begin, we will review some basics about the administrative claims data that we will use for our utilization work.

Review and Approach

Review: Basics of Medicare Claims Data

In Chapter 2 through Chapter 5, we discussed the following characteristics of programming with Medicare administrative data (note that this review is also relevant for Chapters 8, 9, and 10, although we will not repeat it in those chapters):

- Using Medicare claims data is not intuitive. The primary purpose of the Medicare payment system is not to create data for research, but to adjudicate and pay claims. This fact has implications for using our administrative data files and means we must understand the Medicare program to use Medicare claims data effectively.
- Data exist for claims paid by both Medicare Part A and Part B. The Medicare claims files are provided separately for final action Part B carrier, durable medical equipment (DME), outpatient, inpatient, skilled nursing facility (SNF), home health, and hospice services.
- The Part B carrier data set is most commonly associated with claims filed for services provided in a doctor's office, like a checkup. However, the carrier data set also includes claims submitted by clinical social workers, chiropractors, ambulance services, nurse practitioners and physician assistants. In addition, the file also includes claims for services performed outside of a doctor's

office, such as ambulatory surgical centers, hospitals, and hospital emergency rooms. In general, the carrier claims file is the largest Medicare claims data set used for research purposes.

- The DME file includes paid claims for wheelchairs and walkers, hospital beds, blood glucose monitors and related supplies, canes and crutches, splints, prosthetics, orthotics, respiratory devices like oxygen equipment and related supplies, and dialysis equipment and supplies.
- The outpatient claims data set contains final action claims filed by institutional outpatient providers. Often users think of these providers as hospital outpatient departments. However, the file also includes the claims of other types of institutional outpatient providers, like ambulatory surgical centers, outpatient rehabilitation facilities, rural health clinics, and even community mental health centers.
- The inpatient claims data set contains final action claims submitted by long stay and short stay inpatient hospitals for the reimbursement of their facility costs, including things like room charges and even some drugs provided during a beneficiary's hospital stay.
- The SNF file contains paid claims for skilled nursing and rehabilitative care in a SNF setting through specialists such as registered nurses, physical therapists, occupational therapists, speech pathologists, and audiologists. The purpose of this care is to treat, observe, and manage beneficiaries' conditions, to help beneficiaries leaving an inpatient hospital to improve or maintain their current condition, and to assist beneficiaries in maintaining their independence. Medicare pays for these services for up to 100 days following an inpatient stay of at least 3 days.
- The home health data set contains final action claims filed by Home Health Agencies (HHAs). HHAs are entities that provide skilled professional care in a beneficiary's home. These agencies provide services to Medicare beneficiaries like occupational therapy, physical therapy, skilled nursing care, and even speech therapy and medical social services like counseling that can help a beneficiary cope with the impacts of their illness on their mental health.
- The hospice claims file contains final action claims data that have been submitted by hospice providers. Hospice programs provide care for Medicare beneficiaries who are terminally ill. Hospice services include physical care provided by doctors and nurses, care provided by hospice aides, occupational therapy, physical therapy, medical equipment and supplies, drugs, and even grief counseling and respite care (care provided in a facility designed to give family members a break from giving care) for the beneficiary's family members. All hospice services are covered as long as they are related to the beneficiary's terminal illness (services not related to the terminal illness are covered by other Medicare benefits). Many of these services may involve pain management, and services may take place in a facility or the patient's home. Hospice care is insured in benefit periods, meaning that a Medicare beneficiary can get hospice care for two 90-day benefit periods and then an unlimited number of 60-day benefit periods.
- At a minimum, the following data elements are commonly required for research programming: Information that identifies the beneficiary, such as a **BENE_ID**; information about the provider of the service, like the NPI; information that describes when the services occurred (start and end dates); codes that describe the beneficiary (medical diagnosis codes); codes that describe the services (like procedure codes); and information on the amount paid for the services.

- It is prudent to plan effectively prior to beginning your programming project. Part of this planning effort involves creating written specifications that will guide the creation of your SAS algorithms. In addition, it is equally important to execute quality assurance and quality control steps throughout your programming process. These QA/QC procedures include reviewing the written specifications and debugging your SAS code through viewing output and test cases, as well as benchmarking output. Finally, because Medicare administrative data files can be quite large, it is important to keep efficient programming techniques in mind when coding.

Our Programming Plan

In Chapter 5, we submitted a data request to extract all claims and enrollment information for beneficiaries who were treated by the providers in our study population. We received and loaded our data, kept only those variables we wished to use in our study, and created claim-level files that contain both the base claim and revenue center (or claim line) detail in a single record for each claim. In Chapter 6, we used enrollment data to delimit our population to full-year FFS beneficiaries. Now, we wish to bring our claims data to bear on the calculation of utilization measurements. To this end, we will begin our exercise by merging our claims data with the beneficiary population outputted in Chapter 6, eliminating claims for those beneficiaries we are not studying due to their enrollment characteristics. Performing this step early on in our process will reduce the size of the claims files, thereby reducing the execution times of our programs. Once we delimit our claims files, we will begin to research and code our utilization algorithms. We will approach our investigation of utilization using a couple of methods. For services that are defined using procedure codes, we will simply count the unique occurrence of those procedure codes. For example, we can identify E&M services by keying on precise procedure codes. For services identified by the presence of a claim on a unique date or span of dates, we will simply count services by unique start and end dates. For example, we will get a handle on SNF utilization by looking at the start and end dates on claims. While we focus on identifying utilization of services using procedure codes or start and end dates on claims, our SAS code can be easily applied to studying utilization using revenue center codes, or to counting instances of a set of diagnosis codes. What's more, while we do not explore utilization of durable medical equipment (DME), we could easily apply the code in this book to studying the utilization of DME, like hospital beds and wheelchairs.

The output of the exercises in this chapter will be a set of analytic files at the beneficiary-level used for the remainder of our programming work. Looking forward, we will continue to build on the work performed in Chapters 5 and 6 (and this chapter). In Chapter 8, we will use the output of this chapter in our study of Medicare payment. In Chapter 9, we will combine the study of chronic conditions with the study of quality of care by identifying services that are provided to beneficiaries with diabetes and COPD. As such, we will add an important dimension to our analysis. In Chapter 10, we will end our example project by summarizing our output.

Algorithms: Delimit Claims by Continuously Enrolled Beneficiary Population

In Step 7.1, we eliminate claims data for beneficiaries outside of our study population of continuously enrolled beneficiaries. We execute this task by performing a single DATA step merge of the claims files loaded in Chapter 5 (except for the DME file, which we do not use for our example research programming project) and the relevant enrollment data (ENR.CONTENR_2010_FNL) created in Chapter 6. Because we wish to eliminate all claims for those beneficiaries who are not in our study population, we must merge the data sets by the beneficiary identifier variable, **BENE_ID**. In order to make our algorithm run as quickly as possible, we want to be explicit about reading only the **BENE_ID** into memory when using our beneficiary population data set. In addition, because we can use the same code to delimit each of our seven claims data sets, we code a simple macro called CLM_LOOP that specifies the name of the input claims data set that we are delimiting. Using this macro allows us to avoid coding the below merging algorithms an additional six times. Reducing the total lines of code we have to write saves development time and reduces the potential for coding errors. Note that we perform an "if a" merge, keeping only those records in a given claims data set that match the **BENE_ID** in our enrollment data.

```
/* STEP 7.1: DELIMIT CLAIMS DATA BY BENEFICIARY POPULATION */
/* CREATE MACRO LOOP TO PROCESS ALL SEVEN CLAIM TYPES */
%macro clm_loop(clmtyp= );
data utl.&clmtyp._2010_fnl;
    merge enr.contenr_2010_fnl(keep=bene_id in=a)
etl.&clmtyp._2010(in=b);
    by bene_id;
    if a;
run;
%mend clm_loop;
%clm_loop(clmtyp=carr);
%clm_loop(clmtyp=dm)
%clm_loop(clmtyp=op);
%clm_loop(clmtyp=ip);
%clm_loop(clmtyp=sn);
%clm_loop(clmtyp=hh);
%clm_loop(clmtyp=hs)
```

Algorithms: Measuring Evaluation and Management (E&M) Utilization

Let's start our investigation of utilization characteristics for our population of continuously enrolled beneficiaries by measuring the number of E&M services in a physician office setting in 2010. In Step 7.2, we create a macro variable called emcodes that identifies the E&M services we are measuring by looking for certain procedure code[1] values beginning with "99" (indicating the "E&M" part of the criteria). We include the following procedure codes in the macro variable emcodes: 99201, 99202, 99203, 99204, 99205, 99211, 99212, 99213, 99214, and 99215. You can learn more about the definitions of these procedure code values online, or by purchasing a procedure code terminology manual. Note that we do not use every procedure code that begins with "99". For example, procedure code values like 99241 and 99245 identify consultations, but those consultations take place outside of a physician's office.

```
/* STEP 7.2: DEFINE E&M CODES */
%let emcodes='99201','99202','99203','99204','99205','99211',
'99212','99213','99214','99215';
```

Next, in Step 7.3, we search the carrier claims data set (the "physician office setting" part of the criteria) UTL.CARR_2010_FNL for the procedure codes contained in our macro variable emcodes. The carrier claims data has 13 line items, so our data set has 13 procedure codes to search on a per-claim (or per-record) basis. We perform our search by creating an array called HCPCSCD that represents the 13 procedure code variables (**HCPCS_CD1-HCPCS_CD13**). Then, we create a flag variable called **EM_SVC** that is set to a value of 1 if the substring of one of the procedure code variables in the array is in the set of procedure codes defined by the macro variable emcodes.[2] If such evidence is found, the leave statement ends the processing, and the next record is searched (in other words, we are not counting all E&M services that could be on the claim, but simply searching for evidence of a single E&M service). The outputted data (EM_UTIL) contains the claims in the carrier data set, with a reduced set of variables that define the beneficiary, provider, dates of service, provider, and flag indicating an E&M service was discovered. This output file contains only those records where **EM_SVC** is equal to 1.

```
/* STEP 7.3: PULL RECORDS FROM CARRIER FILE BASED ON EMCODES LIST */
data em_util(keep=bene_id em_svc);
        set utl.carr_2010_fnl (keep=bene_id expnsdt: prfnpi: hcpcs_cd:);
        array hcpcscd(13) hcpcs_cd1-hcpcs_cd13;
        em_svc=0;
        do i=1 to 13;
                if substrn(hcpcscd(i),1,5) in(&emcodes) then do;
                        em_svc=1;
                        leave;
                end;
        end;
        if em_svc=1;
run;
```

In Step 7.4, we de-duplicate the file EM_UTIL by beneficiary identifier, keeping only one record per beneficiary, where there is evidence of an E&M service. We output a permanent file called UTL.EM_UTIL. In other words, this file is a list of beneficiaries for whom we found evidence of an E&M service in our carrier claims data.

```
/* STEP 7.4: DE-DUPLICATE TO CREATE LIST OF BENES WITH AT LEAST ONE E&M
SERVICE */
proc sort data=em_util out=utl.em_util nodupkey;
        by bene_id;
run;
```

Algorithms: Measuring Inpatient Hospital Utilization

Our inpatient data are at the claim level, not the stay level. An inpatient stay can be comprised of many claims. CMS offers a stay-level file, with these claims collapsed to a single stay record, but we did not request this data set. On past projects, I have been responsible for the complicated process of writing specifications to convert inpatient claims into stay-level records. Luckily, stays comprised of multiple claim records are not common. Indeed, stays comprised of multiple claims comprise only one or two percent of all inpatient claims. Therefore, at the risk of slightly over-counting the number of stays in our inpatient data, we will proceed with our discussion of stays as follows:

1. To keep the book at the introductory level, we will identify and keep only those claims in our inpatient data set for acute short-stay hospitals (more on this below).
2. We will treat our inpatient claim records as stays and ignore the fact that we have a very small number of hospital stays in our inpatient data that are comprised of multiple claims. We will also ignore the fact that some records represent transfers to other hospitals.[3] (See the Exercises section below for a question on how to collapse multiple claims stays into a single record.) We use these claims to calculate the number of inpatient stays, the length of each stay, and the average length of stay for a beneficiary.

First, in Step 7.5, we focus on delimiting our population to acute short-stay hospitals, as well as keeping only those records with non-negative payment amounts. Medicare distinguishes acute short-stay hospitals between those that are subject to the Medicare Inpatient Prospective Payment System (IPPS) and critical access hospitals (CAHs). IPPS hospitals are paid an amount set prospectively (i.e., determined in advance) based on a patient's condition (i.e., diagnosis) and treatment (i.e., procedures). These payments are based on the Medicare Severity Diagnosis-Related Groups (MS-DRGs). CAHs are limited to 25 or fewer beds and primarily operate in rural areas. They are paid on the basis of cost (i.e., not prospectively). We delimit our inpatient claims data set to these acute short-stay hospitals because these hospitals are seen as different from hospitals like inpatient rehabilitation, long-term care, and inpatient psychiatric facilities (e.g., CMS's Hospital Compare[4] only reports on acute short-stay hospitals).

```
/* STEP 7.5: IDENTIFY AND KEEP ONLY THOSE CLAIMS FOR SHORT STAY
HOSPITALS AND NON-NEGATIVE PAYMENTS */
data utl.ip_2010_fnl_ss;
      set utl.ip_2010_fnl;
      provtype = substr(provider,3,4);
      if (('0001'<=provtype<='0899') or ('1300'<=provtype<='1399')) and
pmt_amt>0;
run;
```

How do we define acute short-stay hospitals? The **PROVIDER** variable is the provider identifier for the hospital and it contains embedded intelligence. Specifically, the first two positions of the **PROVIDER** variable identify the SSA state code for the hospital's location, and the third and fourth positions identify the category of the hospital. We will use this category identifier to find IPPS hospitals and CAHs. IPPS hospitals are identified where the substring of the **PROVIDER** variable, starting at the third position through the sixth position, is equal to 0001 through 0899. CAHs are identified where the substring of the **PROVIDER** variable, starting at the third position through the sixth position, is equal to 1300 through 1399. Therefore, we create a variable called **PROVTYPE** that is this substring of the **PROVIDER** variable, starting at the third position and ending with the sixth position (inclusive). Then, we output a data set called UTL.IP_2010_FNL_SS, keeping only those records with values of **PROVTYPE** that identify IPPS and critical access hospitals. We output this data set to a permanent library because we will use it again in Chapter 8 when we measure Medicare payments for inpatient services (i.e., we will measure payments only for those services that occurred in acute short-stay hospitals).

In addition to keeping only those inpatient claims that occurred in acute short-stay hospitals, our code in Step 7.5 also uses the Medicare payment amount variable (**PMT_AMT**) to remove inpatient claims with negative payment amounts. There should be very few such claims. Generally, these claims are excluded because the negative Medicare payment amount is seen as an indicator that some information on the claim may not be reliable. In some instances, investigators do not use claims with negative payment amounts for payment analysis, but do use these claims for utilization and quality analyses. For example, in studying quality of care, we may choose to "err on the side of the provider" and look for evidence of a service that indicates high quality care (like a retinal eye exam) regardless of the value of payment amount. In our example research programming project, we will keep it simple and exclude these records from the inpatient data for the remainder of our work. Therefore, from this point forward, we will use the output of Step 7.4, called UTL.IP_2010_FNL_SS, for all work requiring inpatient claims data.

In Step 7.6, we use the UTL.IP_2010_FNL_SS data set that contains only claims for acute short-stay hospitals to calculate the length of stay (**IP_LOS**), measured in days for each beneficiary's hospital stay. In simple terms, the length of stay is the discharge date (**DSCHRGDT**) less the admission date (**ADMSN_DT**), plus one day to account for cases where the admission and discharge dates fall on the same day (without adding one day, the result would be a stay of zero days). In some cases, you may consider writing separate algorithms for stays where the admission and discharge dates are equal, and for stays lasting longer than a single day. We also perform a frequency distribution of the length of stay calculations. We output a data set called STAY_UTIL that contains beneficiary and provider identifiers, the start and end dates of the hospitalization, and the calculated length of stay variable. We also perform a frequency distribution of **IP_LOS**, using SAS' Output Delivery System to create our results.

```
/* STEP 7.6: CALCULATE LOS, ASSUMING EVERY INPATIENT CLAIM IS A STAY */
data stay_util(keep=bene_id admsn_dt dschrgdt provider ip_los);
      set utl.ip_2010_fnl_ss;
      ip_los=dschrgdt-admsn_dt+1;
run;

ods html file="C:\Users\mgillingham\Desktop\SAS
Book\FINAL DATA\ODS_OUTPUT\Gillingham_fig7_6.html"
image_dpi=300 style=GrayscalePrinter;
ods graphics on / imagefmt=png;
title "FREQ OF INPATIENT LENGTH OF STAY";
proc freq data=stay_util(where=(1<=ip_los<=10));
      tables ip_los;
run;
ods html close;
```

Output 7.1 shows the results of Step 7.6, delimited for display purposes to count only those records where the IP_LOS is between 1 and 10 days, inclusive.

Output 7.1: Frequency of Inpatient Length of Stay

FREQ OF INPATIENT LENGTH OF STAY

The FREQ Procedure

ip_los	Frequency	Percent	Cumulative Frequency	Cumulative Percent
1	126	1.86	126	1.86
2	938	13.85	1064	15.71
3	1185	17.50	2249	33.22
4	1318	19.47	3567	52.68
5	930	13.74	4497	66.42
6	729	10.77	5226	77.18
7	575	8.49	5801	85.67
8	417	6.16	6218	91.83
9	325	4.80	6543	96.63
10	228	3.37	6771	100.00

In Step 7.7, we modify the STAY_UTIL data set by calculating the number of stays for each beneficiary, as well as the total number of days spent in the hospital for the year (**TOT_IP_LOS**). We use a PROC SQL statement that creates a variable called **STAY_CNT** that is the count of the number of claims, grouped by **BENE_ID** (beneficiary), in our input data set. At the same time, we add the IP_LOS values for each beneficiary to create the **TOT_IP_LOS** variable for use in calculating the average length of stay in the next step. We output a data set called UTL.STAY_UTL. Because we used **distinct(BENE_ID)** in the select statement, our output contains one record per beneficiary (so the **IP_LOS** variable is no longer relevant).

```
/* STEP 7.7: CALCULATE NUMBER OF STAYS AND TOTAL IP LOS, ASSUMING EVERY
INPATIENT CLAIM IS A STAY */
proc sql;
    create table utl.stay_util as
       select distinct(bene_id), admsn_dt, dschrgdt, count(bene_id) as
stay_cnt, sum(ip_los) as tot_ip_los
       from stay_util
       group by bene_id;
quit;
```

In Step 7.8, we modify the UTL.STAY_UTIL data set by calculating the average length of stay (**IP_ALOS**), measured in days for each beneficiary. For a provider, the ALOS statistic provides information on how quickly patients are being discharged after treatment. For a beneficiary, the ALOS statistic provides information on the average length of hospital stays. We calculate ALOS by dividing **IP_LOS** by **STAY_CNT** and performing a frequency distribution of **IP_ALOS**. Again, we use ODS to output the results of the frequency distribution.

```
/* STEP 7.8: CALCULATE ALOS */
data utl.stay_util;
       set utl.stay_util;
       ip_alos=tot_ip_los/stay_cnt;
run;

ods html file="C:\Users\mgillingham\Desktop\SAS
Book\FINAL_DATA\ODS_OUTPUT\Gillingham_fig7_8.html"
image_dpi=300 style=GrayscalePrinter;
ods graphics on / imagefmt=png;
title "FREQ OF INPATIENT AVG LENGTH OF STAY";
proc freq data=utl.stay_util(where=(1<=ip_alos<=5));
       tables ip_alos;
run;
ods html close;
```

Output 7.2 shows the results of Step 7.8, delimited to include only those records with a value of IP_ALOS between 1 and 5 days, inclusive.

Output 7.2: Frequency of Inpatient Average Length of Stay

FREQ OF INPATIENT AVG LENGTH OF STAY

The FREQ Procedure

ip_alos	Frequency	Percent	Cumulative Frequency	Cumulative Percent
1	99	2.33	99	2.33
1.5	6	0.14	105	2.47
2	786	18.49	891	20.95
2.3333333333	3	0.07	894	21.03
2.5	40	0.94	934	21.97
3	1025	24.11	1959	46.07
3.3333333333	9	0.21	1968	46.28
3.5	102	2.40	2070	48.68
3.6	5	0.12	2075	48.80
3.6666666667	18	0.42	2093	49.22
4	1146	26.95	3239	76.18
4.3333333333	12	0.28	3251	76.46
4.5	130	3.06	3381	79.52
4.6666666667	9	0.21	3390	79.73
5	862	20.27	4252	100.00

Algorithms: Measuring Emergency Department Utilization

As discussed in Chapter 3, the identification of emergency department utilization is an interesting and challenging task. Proper identification of emergency room services involved the use of three claims data sets: inpatient data, outpatient data, and Part B carrier data.[5] Specifically:

- If a beneficiary is admitted to the hospital following the ED visit, the information pertaining to the ED visit can be found in the inpatient data. These ED services are identified using revenue center codes 0450-0459 and 0981.
- If a beneficiary is seen in an ED but is not subsequently admitted to the hospital, the claim for the ED visit can be found in the outpatient data. These claims can be identified using the same revenue center codes used for identification of ED visits in the inpatient data (revenue center codes 0450-0459 and 0981).
- The professional component for ED services may appear in the Part B carrier data. This professional component can be identified using certain procedure code values (such as 99281, 99282, 99283, 99284, and 99285). We could also use (or add) a place of service code value of 23.

If we were measuring total ED costs, or counting total ED visits, we would identify ED services using all three claims data sets. However, because we are studying a set of providers, we will focus on identifying the professional component of ED services. In Step 7.9, we use the carrier file created in Step 7.1 above (UTL.CARR_2010_FNL) to identify the professional component of ED services. Our code is similar to the code used in Step 7.3, creating an array for the 13 HCPCS codes (**HCPCS_CD1-HCPCS_CD13**) on carrier claims, and a flag representing an ED service (**ED_SVC**). We search each claim for procedure codes equal to 99281, 99282, 99283, 99284, and 99285. These procedure code values indicate emergency department services (a procedure code value of 99288 may be considered because it indicates direction by a physician of emergency medical systems care during the transportation of the patient to the emergency department, but Medicare does not pay it).[6] We output a file called UTL.ED_CARR_UTIL.

```
/* STEP 7.9: IDENTIFY THE PROFESSIONAL COMPONENT OF ED SERVICES */
data utl.ed_carr_util(keep=bene_id ed_svc);
        set utl.carr_2010_fnl;
        array hcpcscd(13) hcpcs_cd1-hcpcs_cd13;
        ed_svc=0;
        do i=1 to 13;
                if substrn(hcpcscd(i),1,5)
in('99281','99282','99283','99284','99285') then do;
                        ed_svc=1;
                        leave;
                end;
        end;
run;

ods html file="C:\Users\mgillingham\Desktop\SAS
Book\FINAL_DATA\ODS_OUTPUT\Gillingham_fig7_9.html"
image_dpi=300 style=GrayscalePrinter;
ods graphics on / imagefmt=png;
```

```
title "PRINT OF ED PROFESSIONAL COMPONENT SERVICES";
proc print data=utl.ed_carr_util(obs=10);
       where ed_svc=1;
run;
ods html close;
```

Finally, we use ODS to display the output of Step 7.9, keeping only those records where **ED_SVC** is equal to 1. Output 7.3 illustrates just the first 10 observations for display purposes.

Output 7.3: ED Professional Component Services

PRINT OF ED PROFESSIONAL COMPONENT SERVICES

Obs	BENE_ID	ed_svc
62	00052705243EA128	1
129	000B97BA2314E971	1
167	00139C345A104F72	1
176	0013E139F1F37264	1
187	0014FFD71C90B753	1
286	0018BD6F2F493452	1
355	001E32373E05BA96	1
361	001E32373E05BA96	1
395	001EA2F4DB30F105	1
490	00240BD4B2B5FF81	1

Algorithms: Measuring Utilization of Ambulance Services

Let's further explore the carrier file by studying the utilization of ambulance services. Checking with a subject matter expert in identifying services using billing codes reveals that we can identify these services using procedure code values of A0021, A0080, A0090, A0100, A0110, A0120, A0130, A0140, A0160, A0170, A0180, A0190, A0200, A0210, A0225, A0380, A0382, A0384, A0390, A0392,

A0394, A0396, A0398, A0420, A0422, A0424, A0425, A0426, A0427, A0428, A0429, A0430, A0431, A0432, A0433, A0434, A0435, A0436, A0800, A0888, and A0999.

Because it is possible for more than one ambulance service to appear on a claim, we use a variant of the code developed in Step 7.3 to identify the occurrence of an E&M visit. Specifically, we must change the code to count the total number of occurrences of a relevant procedure code, instead of using a leave statement to end the processing after a single occurrence of a relevant procedure code is found.

In Step 7.10, we create a macro variable called ambcodes that identifies the procedure code values we are looking for. Then, we use the Part B carrier data set created in Step 7.1 (UTL.CARR_2010_FNL) to search for the procedure codes represented by the ambcodes macro variable. We create an array for the 13 procedure codes (HCPCS_CD1-HCPCS_CD13) on the carrier claims, as well as a variable called AMB_CARR_SVC used to count the total number of relevant procedure codes on the claim. Note the absence of a leave statement that ends the processing when a relevant value of the procedure code variable is found. In place of the leave statement, we include logic that iterates the value of AMB_CARR_SVC on a record by one each time a relevant procedure code (i.e., one of the procedure codes represented by ambcodes) is found on that record. We output a file called AMB_UTIL, keeping only the BENE_ID, procedure codes, and the AMB_CARR_SVC variables.

```
/* STEP 7.10: IDENTIFY AMBULANCE SERVICES IN CARRIER CLAIMS */
/* DEFINE AMBULANCE PROCEDURE CODES */
%let ambcodes='A0021','A0080','A0090','A0100','A0110',
'A0120','A0130','A0140','A0160','A0170','A0180','A0190','A0200','A0210',
'A0225','A0380','A0382','A0384','A0390','A0392','A0394','A0396','A0398',
'A0420','A0422','A0424','A0425','A0426','A0427','A0428','A0429','A0430',
'A0431','A0432','A0433','A0434','A0435','A0436','A0800','A0888','A0999';

data amb_util(keep=bene_id hcpcs_cd: amb_carr_svc);
       set utl.carr_2010_fnl;
       array hcpcscd(13) hcpcs_cd1-hcpcs_cd13;
       retain amb_carr_svc;
       amb_carr_svc=0;
       do i=1 to 13;
              if substrn(hcpcscd(i),1,5) in(&ambcodes) then do;
                     amb_carr_svc=amb_carr_svc+1;
              end;
       end;
       if amb_carr_svc>=1;
run;

ods html file="C:\Users\mgillingham\Desktop\SAS
Book\FINAL_DATA\ODS_OUTPUT\Gillingham_fig7_10.html"
image_dpi=300 style=GrayscalePrinter;
ods graphics on / imagefmt=png;
title "PRINT OF AMBULANCE SERVICES BY BENEFICIARY";
proc print data=amb_util(where=(bene_id='0073EAD53F4BBC1C')); run;
ods html close;
```

Here is the AMB_UTL data for a single beneficiary who has more than one claim for ambulance services, displayed using ODS. You can see that the value of **AMB_CARR_SVC** matches the number of relevant procedure codes on the claim.

PRINT OF AMBULANCE SERVICES BY BENEFICIARY

Obs	BENE_ID	hcpcs_cd1	hcpcs_cd2	hcpcs_cd3	hcpcs_cd4	hcpcs_cd5	hcpcs_cd6	hcpcs_cd7	hcpcs_cd8	hcpcs_cd9	hcpcs_cd10	hcpcs_cd11	hcpcs_cd12	hcpcs_cd13	amb_carr_svc
81	0073EAD53F4BBC1C	99309	A0425	85027	83002		86376		84480	86376	83550	87340	82607	86803	1
82	0073EAD53F4BBC1C	99232	A0425	85027	83002		86376		84480	86376	83550	87340	82607	86803	1
83	0073EAD53F4BBC1C	82108	A0425	85027	83002		86376		84480	86376	83550	87340	82607	86803	1
84	0073EAD53F4BBC1C	97750	A0425	85027	83002		86376		84480	86376	83550	87340	82607	86803	1
85	0073EAD53F4BBC1C	70450	A0425	85027	83002		86376		84480	86376	83550	87340	82607	86803	1
86	0073EAD53F4BBC1C	90960	A0425	85027	83002		86376		84480	86376	83550	87340	82607	86803	1
87	0073EAD53F4BBC1C	64510	A0425	85027	83002		86376		84480	86376	83550	87340	82607	86803	1
88	0073EAD53F4BBC1C	A0428	99232	99232	99232		86376		84480	86376	83550	87340	82607	86803	1
89	0073EAD53F4BBC1C	99304	A0425	4177F	99232		86376		84480	86376	83550	87340	82607	86803	1
90	0073EAD53F4BBC1C	A0428	A0425	J9035	99232		86376		84480	86376	83550	87340	82607	86803	2

In Step 7.11 we count the total number of ambulance services per beneficiary. Specifically, we sum the values of **AMB_CARR_SVC** by beneficiary. The output of this summation is stored in a variable called **TOT_AMB**. We output a permanent file called UTL.CNT_SVCS_AMB_UTIL, keeping only one record per distinct value of **BENE_ID**.

```
/* STEP 7.11: COUNT AMBULANCE SERVICES PER BENEFICIARY */
proc sql;
        create table utl.amb_util as
        select distinct(bene_id), sum(amb_carr_svc) as tot_amb_svc
        from amb_util
        group by bene_id;
quit;

ods html file="C:\Users\mgillingham\Desktop\SAS
Book\FINAL_DATA\ODS_OUTPUT\Gillingham_fig7_11.html"
image_dpi=300 style=GrayscalePrinter;
ods graphics on / imagefmt=png;
title "PRINT OF TOTAL AMBULANCE SERVICES COUNTS BY BENEFICIARY";
proc print data=utl.amb_util(where=(bene_id='0073EAD53F4BBC1C')); run;
ods html close;
```

Output 7.4 shows the UTL.CNT_SVCS_AMB_UTIL data for the same beneficiary displayed above. You can see that the value of **TOT_AMB** is indeed the summation of the values of **AMB_CARR_SVC** displayed above.

Output 7.4: Total Ambulance Services Counts

PRINT OF TOTAL AMBULANCE SERVICES COUNTS BY BENEFICIARY

Obs	BENE_ID	tot_amb_svc
41	0073EAD53F4BBC1C	11

Algorithms: Measuring Outpatient Visit Information

In Step 7.12, we present a simple formula for calculating the number of outpatient visits for each beneficiary. We treat each record in the outpatient file as a claim for a distinct visit (making assumptions similar to those for inpatient data, and for similar reasons, and with similar implications). Therefore, we use PROC SQL to count each claim for a beneficiary (creating a variable called **OP_VISIT_CNT** using a group by statement). We output a data set called UTL.OP_UTIL, and perform a PROC PRINT to display the summarized information contained therein, using ODS to create our results.

```
/* STEP 7.12: MEASURE OUTPATIENT UTILIZATION */
proc sql;
    create table utl.op_util as
       select distinct(bene_id), count(bene_id) as op_visit_cnt
       from utl.op_2010_fnl
       group by bene_id;
quit;

ods html file="C:\Users\mgillingham\Desktop\SAS
Book\FINAL_DATA\ODS_OUTPUT\Gillingham_fig7_12.html"
image_dpi=300 style=GrayscalePrinter;
ods graphics on / imagefmt=png;
title "PRINT OF OUTPATIENT VISIT SUMMARIES BY BENEFICIARY";
proc print data=utl.op_util(obs=10); run;
ods html close;
```

Output 7.5 shows the results of Step 7.12, printing just the first 10 observations for display purposes.

Output 7.5: Outpatient Visit Summaries

PRINT OF OUTPATIENT VISIT SUMMARIES BY BENEFICIARY

Obs	BENE_ID	op_visit_cnt
1	00016F745862898F	1
2	0001FDD721E223DC	1
3	00021CA6FF03E670	1
4	0002DAE1C81CC70D	1
5	000308435E3E5B76	1
6	000489E7EAAD463F	1
7	0004F0ABD505251D	1
8	00052705243EA128	1
9	00070B63745BE497	1
10	0007F12A492FD25D	7

Algorithms: Measuring Utilization of SNF, Home Health Agency, and Hospice Care

Claims for SNF and hospice services are billed in 30-day increments; claims for home health services are billed in 60-day increments. If a beneficiary is discharged (or expires) prior to the completion of a 30- or 60-day period, the last claim in a set may be for less that 30 or 60 days, respectively. Therefore, when sorting claims by **FROM_DT** and **THRU_DT** for a beneficiary's home health care, we expect to see some claims with a length of stay of 60 days. We may also see claims with a length of stay of less than 60 days for the last claim in the stay.[7] In order to determine length of stay, we must collapse claims that comprise a single stay.

- Scenario #1: Claims with consecutive dates of service, defined where the **THRU_DT** on the earlier claim is within one day of the **FROM_DT** on the next earliest claim. For example, the first claim in a set contains a **FROM_DT** equal to March 1, 2010 and a **THRU_DT** equal to April 30, 2010, and the next consecutive claim in the set has a **FROM_DT** equal to May 1, 2010 and a **THRU_DT** equal to June 30, 2010.

- Scenario #2: Claims with a **FROM_DT** that falls before the **THRU_DT** on another claim in the set. For example, if one claim contains a **FROM_DT** equal to March 1, 2010 and a **THRU_DT** equal to April 30, 2010, and another claim has a **FROM_DT** equal to April 25, 2010 and a **THRU_DT** equal to June 24, 2010.
- Scenario #3: Claims with a **FROM_DT** and **THRU_DT** that fall within the **FROM_DT** and **THRU_DT** of another claim for the beneficiary. For example, one claim contains a **FROM_DT** equal to March 1, 2010 and a **THRU_DT** equal to April 30, 2010, and another claim contains a **FROM_DT** equal to March 5, 2010 and a **THRU_DT** equal to April 4, 2010.
- Scenario #4: Claims that do not fall into any of the above categories. For example, one claim contains a **FROM_DT** equal to March 1, 2010 and a **THRU_DT** equal to April 30, 2010, and the next consecutive claim in the set has a **FROM_DT** equal to July 1, 2010 and a **THRU_DT** equal to August 30, 2010.

In all cases, we want to collapse sets of claims into a single record such that the **FROM_DT** and **THRU_DT** on the collapsed record are equal to the earliest **FROM_DT** in the set and the latest **THRU_DT** in the set, respectively. In Step 7.13 through Step 7.15, we present an algorithm to collapse beneficiary's claims into a single record where appropriate.[8] Our algorithm utilizes a macro called CLM_LOOP2 that is very similar to the CLM_LOOP macro used in Step 7.1 above (the difference being that the CLM_LOOP2 macro only processes the SNF, home health, and hospice data sets). First, in Step 7.13, we sort the relevant claims data set created in Step 7.1, outputting intermediate data sets to our work library. Note that we only keep the beneficiary identifier, **FROM_DT**, and **THRU_DT** on the claim because those are the only variables needed for our work. The sort key (**BENE_ID**, **FROM_DT**, and **THRU_DT**) orders each beneficiary's claims by date in preparation for the processing performed in later steps.

```
/* STEP 7.13: SORT FILE BY BENE_ID, FROM_DT, THRU_DT */
%macro clm_loop2(clmtyp=);
proc sort data=utl.&clmtyp._2010_fnl out=&clmtyp._2010_fnl(keep=bene_id
from_dt thru_dt);
        by bene_id from_dt thru_dt;
run;
```

Step 7.14 uses a lag function to create a variable called **CLAIM_SEGMENT_ID** to use for identifying claims to collapse. The key to understanding this step is to recognize that the 'else do' statement executes first for new combinations of **BENE_ID** and **FROM_DT** values, and **CLAIM_SEGMENT** and **END2** are re-initialized for a new group of overlapping claims. This lag function reads the previous record for a beneficiary (remember that the records are sorted in consecutive order) and determines the relationship of the **THRU_DT** on the previous claim to the **FROM_DT** on the next consecutive claim. If this relationship is described by Scenario #1, Scenario #2, or Scenario #3 above, the **CLAIM_SEGMENT_ID** does not increase by 1, meaning that claims in these scenarios will have the same value of **CLAIM_SEGMENT_ID**. The output of this step is an intermediate data set saved to our work library.

```
/* STEP 7.14: IDENTIFY CLAIMS WITH DATE RANGES TO BE COLLAPSED USING
CLAIM_SEGMENT_ID */
data &clmtyp._2010_stay(drop=bene_id2 thru_dt rename=(from_dt=begin_dos
end2=end_dos));
	set &clmtyp._2010_fnl;
	retain end2;
	bene_id2=lag1(bene_id);
	if bene_id2=bene_id and from_dt le (end2+1) then do;
		from_dt=end2;
		end2=max(thru_dt,end2);
	end;
	else do;
		claim_segment_id+1;
		end2=thru_dt;
	end;
	format end2 mmddyy10.;
run;
```

Step 7.15 keys on the **CLAIM_SEGMENT_ID** to collapse the records outputted in Step 7.14. Specifically, for records with the same value of **CLAIM_SEGMENT_ID**, the code outputs a single record that contains a value of **FROM_DT** equal to the earliest **FROM_DT** in the set of claims, as well as a value of **THRU_DT** equal to the latest value of **THRU_DT** in the set of claims. In the case of records that fall in Scenario #4, the code will output multiple records for the beneficiary. Finally, the code calculates the length of stay (called **LOS**) as the difference of the **THRU_DT** and the **FROM_DT** on each outputted record. The outputted analytic files, UTL.SN_2010_STAY, UTL.HH_2010_STAY, and UTL.HS_2010_STAY, are saved as permanent data sets for future use.

```
/* STEP 7.15: COMBINE CLAIMS WITH OVERLAPPING SEGMENTS */
data utl.&clmtyp._2010_stay_fnl(drop=begin_dos end_dos
claim_segment_id);
	retain from_dt thru_dt;
	set &clmtyp._2010_stay;
	by bene_id claim_segment_id;
	format from_dt thru_dt mmddyy10.;
	if first.claim_segment_id then do;
		from_dt=begin_dos;
	end;
	if last.claim_segment_id then do;
		thru_dt=end_dos;
		los=thru_dt-from_dt;
		output;
	end;
run;
%mend;
%clm_loop2(clmtyp=sn);
%clm_loop2(clmtyp=hh);
%clm_loop2(clmtyp=hs);
```

Chapter Summary

In this chapter, we used our claims and enrollment data to program measurements of utilization and output analytic files containing utilization data for our population. Specifically, we:

- Delimited the claims data we will use throughout the remainder of this book by the beneficiary population we created in Chapter 6.
- Programmed an algorithm to measure utilization of evaluation and management (E&M) services in a physician office setting by defining procedure codes that define E&M visits and searching the array of line-level procedure code variables in the Part B carrier data for any one of those procedure codes.
- Programmed an algorithm to calculate inpatient hospital stays, discussed how most records in the inpatient data set represent a stay, learned how to define acute short-stay hospitals, and programmed an algorithm to determine the length of stay.
- Programmed an algorithm to measure the professional component of emergency department (ED) services by defining how to find ED services in claims data and learning how to search procedure codes for some of those services.
- Programmed an algorithm to measure utilization of ambulance services by defining the procedure codes that characterize ambulance services and counting the total occurrences of those codes on a given claim.
- Programmed algorithms for measures of utilization for outpatient visits, SNFs, home health agencies, and hospice services.

Exercises

1. Can you rearrange programming in Step 7.6 through Step 7.9? Does the program run faster?
3. Can you write code for collapsing stays comprised of multiple claims into a single record containing all information for the stay?
4. Can you think of how the algorithms for identifying the professional component of ED services can be modified to search the inpatient and outpatient files for the existence of revenue center codes that define an ED service?
5. Perform the following thought experiment: Assuming you have data on durable medical equipment claims, apply what you learned in this chapter to write some pseudocode to identify utilization of wheelchairs and hospital beds.
6. Can you apply the logic used to define stays in the Skilled Nursing Facility file to inpatient data to account for multi-claim stays in the inpatient file?

---

[1] As discussed in Chapter 3, Healthcare Common Procedure Coding System (HCPCS) Level I codes, also known as Current Procedural Terminology (CPT) codes, are five digit numeric codes describing medical services, and are a copyrighted coding schema of the American Medical Association (AMA). HCPCS Level II codes are five character alphanumeric codes defining services not described by the Level I codes.

[2] Note that the use of the substring here is a personal preference and may not be necessary in your application.

[3] If you wish to look into transfers further, they can be simply defined by identifying two or more sequential claims for a beneficiary, where the providers are different, and the first date of service on the latter claim (represented by the **ADMSN_DT** variable) is equal to the last date of service or one day after the last date of service on the previous claim (represented by the **DSCHRGDT** variable). For more information, see ResDAC's article called Identifying Multiple Claims for a Hospital or Skilled Nursing Facility Stay, available at http://www.resdac.org/resconnect/articles/125.

[4] The Hospital Compare website is at http://www.medicare.gov/hospitalcompare.

[5] Specifications in this section adapted from ResDAC's article called How to Identify Emergency Room Services in the Medicare Claims Data, available at http://www.resdac.org/resconnect/articles/144.

[6] In general, while there are a number of reasons that beneficiaries are seen in an ED, which could encompass a large range of E&M codes, we are focusing on ED E&M codes as defined in the HCPCS manual.

[7] In addition, the **FROM_DT** is always equal to the **THRU_DT** on the first claim in a set for home health services (see the definition of the **FROM_DT** variable available on ResDAC's website at http://www.resdac.org/cms-data/variables/Claim-Date).

[8] This algorithm is adapted from a SUG paper by Doug Shannon and Wade Bannister, available on VALSUG's website at http://www.valsug.org/meetings/2004/jan/SASPaper-OverlappingDateSegments-D&W.pdf. I like its comprehensive approach to identifying and collapsing records in a SAS data set with consecutive and overlapping dates, as well as records that are not consecutive or overlapping.

Chapter 8: Measuring Costs to Medicare

Introduction and Goals

Measuring cost is important because it gets to the heart of the sustainability of any insurance program, and Medicare is certainly no exception. It is no understatement to say that the measurement of cost in the Medicare program is a hot topic. Open any newspaper or magazine or turn on the evening news and you will see that reducing cost without negatively affecting quality of care is the order of the day. In other words, the mission is to monitor and improve the value of purchased healthcare. To play a part in accomplishing this mission, we must learn how to measure the cost of care. In addition, we must learn to measure the quality of care, which we will do in Chapter 9 as we look at chronic conditions.

Utilization and quality are intimately linked to cost, but not always in ways one might assume. In your research, you may wish to go beyond simple measurements of utilization and cost and actually define categories of services whose payments may have been reduced by an incentive program. For example, did a program reduce the number of hospital stays, thus reducing total cost per beneficiary? Alternatively, did a program reduce the number of annual checkups, resulting in higher costs driven by an increase in the number of hospital stays?

In this text, the term “cost” is limited to the cost of the service to Medicare. We will not study the total cost of services, which include costs to beneficiaries, like coinsurance and deductibles. We are not studying services not covered by Medicare, like cosmetic surgery. In addition, because we retained only claims for acute short-stay hospitals in our inpatient data, we are not including payments for some services, like inpatient rehabilitation. Wrangling with all of these issues is beyond the scope of this introductory text, but it is important for readers to begin to think about these issues and prepare to account for them (or study them!) in their own research work. Rather, we utilize the Medicare payment variables in our claims data sets (i.e., the amount paid by the Medicare program to the provider for the services rendered) to determine costs to Medicare. Therefore, when we use the term "cost," we are speaking about the payment made by Medicare for the service, which is the same thing as the cost of the service to Medicare.

It is worth noting that Medicare does not pay for all services in the same manner. Since 1984, Medicare has been moving away from fee schedule payment and cost reimbursement to Prospective Payment Systems (PPS). Specifically, in very simple terms[1]:

- Carrier claims are paid on a fee schedule.
- DME claims are paid using categorized fee schedules.
- As we discussed in Chapter 7, inpatient claims can be paid prospectively (using the Inpatient Prospective Payment System, or IPPS) or on a cost basis (e.g., for critical access hospitals). Hospitals in Maryland are paid on a rate commission because they do not participate in IPPS.
- Outpatient claims are paid using the Outpatient Prospective Payment System (OPPS). Medicare Part B pays for many of the outpatient services a beneficiary receives. For this reason, the claims in the outpatient data are sometimes referred to as "institutional Part B" claims.
- Skilled nursing facility claims are paid prospectively. The Prospective Payment System (PPS) for SNF claims is based on a predetermined payment rate for each day of care.
- Home health agency claims are paid prospectively. The PPS for home health claims is based on a predetermined payment rate applied to each 60-day episode of home health care.
- Hospice care claims are paid based on a daily payment rate that does not take into account the level of services provided on any given day of hospice care.

In order to get a handle on the cost outcomes of the incentive program, we will code algorithms to do the following:

- Measure evaluation and management (E&M) costs in a physician office setting.
- Measure inpatient hospital costs.
- Measure total Part A costs (i.e., those costs associated with inpatient, skilled nursing facility (SNF), home health, and hospice services).

You may have noted that we are not looking at ED costs. Why? As we discussed in Chapter 7, we use separate criteria for identifying emergency department services that resulted in an admission to the hospital, and those not resulting in an admission to the hospital. Although identifying the professional component of emergency department costs on carrier claims data sets is straightforward, compiling cost information from the outpatient and inpatient claims data is not straightforward, and beyond the scope of this text. For example, emergency department services found in the inpatient file revenue center line items are not necessarily reflective of the payment made by Medicare for those services. Although it is possible to determine these costs, it is not an easy exercise. Therefore, we will not study payments made by Medicare for emergency department services in this chapter.

As discussed in earlier chapters, the online companion to this book is at http://support.sas.com/publishing/authors/gillingham.html. Here, you will find information on creating dummy source data, the code in this and subsequent chapters, as well as answers to the exercises in this book. I expect you to visit the book's website, create your own dummy source data, and run the code yourself.

Review and Approach

Review: Basics of Medicare Cost

In our introduction to Medicare covered in Chapter 2, we discussed the following topics related to Medicare cost:

- Medicare may require beneficiaries to make certain cost-sharing payments, like deductibles and coinsurance.
- Medicare may not be the primary payer for services provided to beneficiaries who carry additional health insurance coverage, even the elderly.
- Medicare is a secondary payer for beneficiaries that have certain additional health insurance coverage, like the Federal Black Lung Program.
- Medicare does not cover every possible medical service or procedure. Some services have limitations on coverage. For example, Medicare Part A stops paying for inpatient psychiatric care in a psychiatric hospital after 190 days (this is a lifetime limit). Other services are simply not covered. For example, Medicare does not cover long-term care services and cosmetic surgery.
- Some limited and uncovered services as well as cost-sharing payments can be covered by supplemental insurance. For example, beneficiaries can acquire supplemental coverage from Medigap insurance policies, insurance sponsored by their employers, Medicare Advantage plans, and, in some cases, Medicaid.

Our Programming Plan

In Chapter 5, we submitted a data request to extract all claims and enrollment information for beneficiaries treated by the providers in our study population. We received and loaded our data, keeping only those variables we wish to use in our study. In Chapter 6, we used enrollment data to delimit our population to full-year FFS beneficiaries. In Chapter 7, we brought our claims data to bear on the calculation of utilization measurements. Now, we will draw on our work in Chapter 7 to study Medicare payments. In some instances, we will simply sum payments by leveraging code that we learned in Chapter 7 to identify specific services. For example, we learned how to flag E&M services in a physician office setting. We only need to leverage this code to identify the relevant services and add simple algorithms to identify and summarize payments. In doing so, we will take advantage of the opportunity to explore different ways to summarize payments and discuss concepts like grouping in PROC SQL.

As in previous chapters, the output of this chapter will be analytic files at the beneficiary level. We will use this output for later programming work. Looking forward, we will continue to build on the work performed in Chapters 5, 6, and 7 (and this chapter). In Chapter 9, we will combine the study of chronic conditions with the study of quality of care by identifying services provided to beneficiaries with diabetes and COPD, adding an important dimension to our analysis. In Chapter 10, we will end our example project by summarizing all of our output.

A Note on Payment Standardization and Risk Adjustment

One important note: In our example project, we plan to compare payments across regions. Two issues can make this effort problematic: Regional variations in cost and variations in cost based on patient severity.

Now, it's time to play "good news and bad news." The good news is that we have solutions to these challenges! The bad news is that the study of these solutions unfortunately falls well outside the scope of this book! Of course, that is not going to stop your fearless author from describing the solutions, called payment standardization and risk adjustment, in a bit more detail!

Payment standardization is a fancy term for controlling for differences in payments that are not related to patient care, such as regional variations in costs that are driven by things like differences in wage rates between communities, standard of living, and the capital costs of hospitals. Risk adjustment is another fancy term that statistically holds constant demographic health status (severity) differences among patients in order to make cost and quality comparisons of medical interventions and treatments.

How does all of this work? If we are trying to draw conclusions about regional variations in costs, it is important to account for the fact that costs in Washington, DC can be higher than those of, say, rural Michigan simply because the cost of living in the Washington, DC metropolitan area is higher. Once we remove the fact that goods and services purchased in Washington, DC cost on average 30% more than those purchased in rural Michigan, we can make accurate comparisons between healthcare costs in these regions.

Likewise, it is important to control for the fact that some providers or groups of providers see patients who are, on average, sicker or older than the patients seen by other providers. In fact, it is easy to imagine a situation where an elderly patient with co-morbidities has a higher chance of complications from surgery than a healthier, younger patient, resulting in the need for more services. This, in turn, may result in higher costs for this beneficiary. Again, both payment standardization and risk adjustment are topics for their own books. We will not use either method in our study, but it is important to know they exist and that CMS has methodologies and models for both subjects.

Algorithms: Measuring Evaluation and Management Payments

We identified evaluation and management services in Chapter 7, but we did not retain the payment variables in our output. What's more, we only searched our carrier claims data for evidence of a single E&M service in a physician office setting for each beneficiary. Now, we want to identify total payments for E&M services in a physician office setting for each beneficiary as well as total and average payments for the population of beneficiaries with at least one E&M service in a physician office setting as a whole. The code we will write is straightforward. We will start out by adding payments to our E&M algorithm that we developed in Chapter 7. Then, we will use PROC SQL to sum payments, and a DATA step to calculate payments per beneficiary.

In Step 8.1, we start with the code we wrote in Chapter 7 to identify E&M services, adding code to keep and use Medicare payment variables (**LINEPMT1-LINEPMT13**) for the summarization of payments at the beneficiary level. In addition, we add code to search for all E&M services on a claim (instead of a single occurrence as was done in Chapter 7).[2] The **LINEPMT** variable defines the Medicare amount paid for the service on a claim line. We add some code to sum the payments on each claim for each E&M service found in the claim lines, creating the variable **EM_COST** (in addition to the variable we created in Chapter 7 called **EM_SVC**).

```
/* STEP 8.1: DEFINE E&M CODES */
%let emcodes='99201','99202','99203','99204','99205','99211',
'99212','99213','99214','99215';

data em_cost(keep=bene_id expnsdt: prfnpi: linepmt: hcpcs_cd: em_svc
em_cost);
        set utl.carr_2010_fnl;
        array hcpcscd(13) hcpcs_cd1-hcpcs_cd13;
        array prvpmt(13) linepmt1-linepmt13;
        retain em_svc em_cost;
        em_svc=0;
        em_cost=0;
        do i=1 to 13;
                if substrn(hcpcscd(i),1,5) in(&emcodes) then do;
                        em_svc=em_svc+1;
                        em_cost=em_cost+prvpmt(i);
                end;
        end;
run;
```

In Step 8.2, we summarize the E&M payments and services derived in Step 8.1 by beneficiary. Specifically, we use PROC SQL code to create a data set called CST.EM_COST_BENE to create two summary variables, EM_TOTCOST_BENE (which summarizes the total payments for E&M services for each beneficiary) and EM_TOTSVC_BENE (which summarizes the total count of E&M services for each beneficiary). This summarization is achieved by grouping the data by BENE_ID. Note that we label both of the variables we create, and we format the EM_TOTCOST_BENE variable to look like a dollar figure.

```
/* STEP 8.2: CALCULATE TOTAL E&M COSTS AND TOTAL E&M SERVICES BY
BENEFICIARY */
proc sql;
        create table cst.em_cost_bene as
        select bene_id, sum(em_cost) as em_totcost_bene format=dollar15.2
label='TOTAL E&M COST FOR THE BENE',
                sum(em_svc) as em_totsvc_bene label='TOTAL E&M SERVICES FOR
THE BENE'
        from em_cost
        group by bene_id
        order by bene_id;
quit;
```

Output 8.1 shows the results of Step 8.2, printing just the first 10 observations for display purposes.

Output 8.1: Total E&M Costs and Total E&M Services

TOTAL E&M COSTS AND TOTAL E&M SERVICES BY BENEFICIARY FOR BENEFICIARIES WITH AT LEAST ONE E&M SERVICE

Obs	BENE_ID	em_totcost_bene	em_totsvc_bene
1	00016F745862898F	$80.00	1
2	0001FDD721E223DC	$0.00	0
3	00021CA6FF03E670	$0.00	0
4	0002DAE1C81CC70D	$30.00	2
5	000308435E3E5B76	$120.00	1
6	000489E7EAAD463F	$0.00	0
7	0004F0ABD505251D	$780.00	5
8	00052705243EA128	$470.00	8
9	00070B63745BE497	$740.00	5
10	0007F12A492FD25D	$870.00	14

Finally, in Step 8.3, we further summarize our E&M payment and service information, but do so only for the population of beneficiaries with at least one claim for an E&M service in a physician office setting, outputting a data set called CST.EM_COST_ALL. We use PROC SQL to create a count of distinct beneficiaries called EM_TOTBENE, and a summary of total payments called EM_TOTCOST. What's more, we perform a calculation using two of our created variables (EM_TOTBENE and EM_TOTCOST) to arrive at the average E&M payment for our population of beneficiaries with at least one E&M service, called EM_AVGCOST. We use a 'where' clause to ensure that our output only takes into account those beneficiaries with at least one E&M service in a physician office setting. We also label the variables we created, and format the payment and average payment calculations to look like dollar figures.

```
/* STEP 8.3: CALCULATE TOTAL AND AVERAGE E&M COSTS FOR ALL BENEFICIARIES
WITH AT LEAST ONE E&M SERVICE */
proc sql;
	create table cst.em_cost_all as
	select count(distinct bene_id) as em_totbene,
sum(em_totcost_bene) as em_totcost format=dollar15.2 label='TOTAL E&M
COST',
		(calculated em_totcost/calculated em_totbene) as em_avgcost
format=dollar10.2 label='AVERAGE BENE E&M COST'
	from cst.em_cost_bene
	where em_totsvc_bene>0;
quit;
```

Output 8.2 shows the result of Step 8.3.

Output 8.2: Calculate Total and Average E&M Costs

CALCULATE TOTAL AND AVERAGE E&M COSTS FOR THE TOTAL POPULATION OF BENEFICIARIES WITH AT LEAST ONE E&M SERVICE

Obs	em_totbene	em_totcost	em_avgcost
1	49759	$14,618,500.00	$293.79

While Step 8.1 and Step 8.2 provide us with total E&M payments for each beneficiary, we can take this opportunity to arrive at the same information slightly differently. In Chapter 5, we created claim files that contained the claim header-level and claim line-level information in a single record. Can we use the line file to search for E&M codes and summarize their payments by beneficiary? Sure! In Step 8.4, we simply search the carrier claim line data for evidence of E&M procedure codes. Then, we summarize the payment on each line by beneficiary using the LINEPMT variable. Note that we could just as easily summarize the information by provider using the PRFNPI variable. Also note that we use a where clause to delimit our computation to HCPCS codes that represent an E&M service.

```
/* STEP 8.4: CALCULATE TOTAL E&M COSTS BY BENEFICIARY USING THE CLAIM
LINE FILE */
proc sql;
	create table cst.em_cost_bene_line as
	select bene_id, sum(linepmt) as em_totcost_bene format=dollar15.2
label='TOTAL E&M COST FOR THE BENE'
	from src.carrier2010line
	where hcpcs_cd in(&emcodes)
	group by bene_id
	order by bene_id;
quit;
```

Output 8.3 shows the results of Step 8.4, printing just the first 10 observations for display purposes. Note that the output of Step 8.4 is different from the output of Step 8.2. In the Exercises section below, we ask why.

Output 8.3: Total E&M Costs by Beneficiary Using the Claim Line File

TOTAL E&M COSTS BY BENEFICIARY USING THE CLAIM LINE FILE

Obs	BENE_ID	em_totcost_bene
1	00016F745862898F	$80.00
2	0002DAE1C81CC70D	$30.00
3	0002F28CE057345B	$910.00
4	000308435E3E5B76	$120.00
5	000345A39D4157C9	$150.00
6	00048EF1F4791C68	$160.00
7	0004F0ABD505251D	$780.00
8	00052705243EA128	$470.00
9	00070B63745BE497	$740.00
10	0007F12A492FD25D	$870.00

Algorithms: Measuring Inpatient Hospital Payments

In Chapter 7, we decided to treat the claims in our inpatient claims data as stays, and we looked at stay-level summary statistics, like average length of stay (ALOS), for short-stay acute care hospitals. We will now use the total amount paid variable at the header level of the claim to calculate the total amount paid by Medicare for a stay at an acute care facility or critical access hospital. This is accomplished with slight modifications to the code we used to calculate E&M payments in Step 8.2 and Step 8.3.

In Step 8.5, we summarize total inpatient payments using the **PMT_AMT** variable[3] in the inpatient claims data set for each beneficiary, outputting a variable called **IP_TOTCOST_BENE** (similar to the way we calculated total E&M payments for each beneficiary in Step 8.2). We output a data set called CST.IP_COST_BENE. Note that this **PMT_AMT** variable is the amount paid by Medicare to the provider for the services listed on the claim. The payment is made from the Medicare Trust Fund.

Occasionally, the reader may note the presence of negative payments in the inpatient claims.[4] Recall that our work in Step 7.5 in Chapter 7 excluded negative payment amounts from the inpatient claims data set.

```
/* STEP 8.5: CALCULATE TOTAL INPATIENT COSTS BY BENEFICIARY FOR
BENEFICIARIES WITH AT LEAST ONE INPATIENT CLAIM */
proc sql;
	create table cst.ip_cost_bene as
	select bene_id, sum(pmt_amt) as ip_totcost_bene format=dollar15.2
label='TOTAL IP COST FOR THE BENE'
	from utl.ip_2010_fnl_ss
	group by bene_id;
quit;
```

Output 8.4 shows the results of Step 8.5, printing just the first 10 observations for display purposes.

Output 8.4: Total Inpatient Costs by Beneficiary

TOTAL INPATIENT COSTS BY BENEFICIARY FOR BENEFICIARIES WITH AT LEAST ONE INPATIENT CLAIM

Obs	BENE_ID	ip_totcost_bene
1	00016F745862898F	$16,000.00
2	0007F12A492FD25D	$43,000.00
3	0013E139F1F37264	$3,000.00
4	00196F0702489342	$15,000.00
5	002354398A00234E	$10,000.00
6	00271F7DF9C2B88A	$28,000.00
7	0028A82FE0CA0802	$9,000.00
8	00292D3DBB23CE44	$3,000.00
9	002B6203FF086ABA	$12,000.00
10	0049CB2A111F2225	$6,000.00

In Step 8.6, we use logic similar to that of Step 8.3 to summarize total inpatient payments for our population of beneficiaries with at least one inpatient claim as a whole, creating a variable called IP_TOTCOST. We also calculate average inpatient payments for this same population (IP_AVGCOST). Our output is saved to a data set called CST.IP_COST_ALL.

```
/* STEP 8.6: CALCULATE TOTAL AND AVERAGE INPATIENT COSTS FOR ALL
BENEFICIARIES WITH AT LEAST ONE INPATIENT CLAIM */
proc sql;
       create table cst.ip_cost_all as
       select count(distinct bene_id) as ip_totbene,
sum(ip_totcost_bene) as ip_totcost format=dollar15.2 label='TOTAL IP
COST',
              (calculated ip_totcost/calculated ip_totbene) as ip_avgcost
format=dollar10.2 label='AVERAGE BENE IP COST'
       from cst.ip_cost_bene;
quit;
```

Output 8.5 shows the result of Step 8.6.

Output 8.5: Total and Average Inpatient Costs for all Beneficiaries

TOTAL AND AVERAGE INPATIENT COSTS FOR ALL BENEFICIARIES WITH AT LEAST ONE INPATIENT CLAIM

Obs	ip_totbene	ip_totcost	ip_avgcost
1	6914	$81,047,310.00	$11,722.20

Algorithms: Measuring Total Part A Payments

In the above work, we have measured payments for specific services (like E&M), or in specific settings (like a physician's office or acute short-stay hospitals). We must continue this work, using the **PMT_AMT** variable to summarize total Medicare Part A payments. To this end, we will summarize the total payments in each of the inpatient, SNF, home health[5], and hospice data sets separately, and then as a whole. Note that we are not including the outpatient data set because, while the outpatient data set contains services performed in an institutional setting, the claims for those services are paid through Medicare Part B. Considering the fact that we are using the same **PMT_AMT** variable present in each data set to summarize payments, we can perform this summary of payments in a standardized fashion by utilizing a macro loop.

First, in Step 8.7, we utilize a macro loop to calculate the total Medicare amount paid by the beneficiary, which outputs a separate file for each claim type. Our macro, called CLM_LOOP, is reminiscent of the macro we utilized in Chapter 7 (Step 7.1) to delimit the data in each claim type by the beneficiaries in our enrollment data, but we added code to conditionally process inpatient data. Specifically, our CLM_LOOP macro includes a simple PROC SQL statement that should look very familiar; it is the same code we used to define other summaries of payments throughout this chapter.

There is, however, one major difference: We must figure out how to process an inpatient claims file that is named differently than the SNF, home health, and hospice files in such a way as to require special coding in the macro. Indeed, the inpatient file we are using has already been processed to exclude negative payment amounts. Therefore, the name of this file does not follow the same pattern as the other Part A files we are processing in the macro. In order to include the processing of the inpatient claims file in the macro, we use %if logic within the macro to identify the type of data set that is being processed.[6] Apart from this, the code uses the claims data created in Chapter 7 (delimited by our enrollment data and, in the case of inpatient claims, type of hospital and value of payment amount), and outputs separate files for each claim type containing beneficiary identifiers (**BENE_ID**) and total payments associated with that beneficiary (**TOTCOST_BENE**). At this point, we have created payment summaries for each of the Part A claim types.

```
/* STEP 8.7: CALCULATE TOTAL BENEFICIARY INSTITUTIONAL COSTS USING MACRO
LOOP */
%macro clm_loop(clmtyp= );
proc sql;
    create table totcost_&clmtyp. as
    select distinct(bene_id), sum(pmt_amt) as totcost_bene
       %if &clmtyp=sn or &clmtyp=hh or &clmtyp=hs %then %do;
             from utl.&clmtyp._2010_fnl
       %end;
       %else %if &clmtyp.=ip %then %do;
             from utl.ip_2010_fnl_ss
       %end;
          group by bene_id;
             quit;
%mend clm_loop;
%clm_loop(clmtyp=ip);
%clm_loop(clmtyp=sn);
%clm_loop(clmtyp=hh);
%clm_loop(clmtyp=hs);
```

Now that we have summarized payments by claim type, we can summarize beneficiaries' total payments for all Medicare Part A services performed during the year. To complete this task, we must sum all of the payment amounts calculated in Step 8.7 by beneficiary. Therefore, Step 8.8 utilizes the output of Step 8.7, concatenating each data set outputted from the macro used in Step 8.7. The concatenation is performed by BENE_ID so that the data set is sorted by beneficiary.7 Next, we leverage the concatenation by beneficiary to perform "by group" processing. Simply, for each beneficiary identifier (BENE_ID), we set the total Part A payment variable (PA_TOTCOST_BENE) equal to 0. Then, we add the payment summary for each beneficiary created in Step 8.7 (the variable TOTCOST_BENE) to the PA_TOTCOST_BENE variable. In this way, the PA_TOTCOST_BENE variable becomes the summary of each beneficiary's Part A payments. When the last record for the beneficiary is reached, the final PA_TOTCOST_BENE value is outputted, and SAS processes data for the next beneficiary. The final output, called CST.PA_TOTCOST_BENE, contains one record for each beneficiary, where PA_TOTCOST_BENE represents the summation of payments from each of the institutional claim types calculated in Step 8.7.

```
/* STEP 8.8: SET COSTS FROM EACH CLAIMS DATA SET AND CALCULATE TOTAL
INSTITUTIONAL COSTS PER BENEFICIARY */
data cst.pa_totcost_bene(keep=bene_id pa_totcost_bene);
        set totcost_ip totcost_sn totcost_hh totcost_hs;
        by bene_id;
        if first.bene_id then pa_totcost_bene=0;
        pa_totcost_bene+totcost_bene;
        if last.bene_id then output;
        format pa_totcost_bene dollar15.2;
        label pa_totcost_bene='TOTAL PART A COSTS PER BENEFICIARY';
run;
```

Output 8.6 shows the results of Step 8.8, just printing the first 10 observations for display purposes.

Output 8.6: Total Part A Costs Per Beneficiary

TOTAL PART A COSTS PER BENEFICIARY

Obs	BENE_ID	pa_totcost_bene
1	00016F745862898F	$64,000.00
2	0001FDD721E223DC	$0.00
3	00021CA6FF03E670	$0.00
4	0002DAE1C81CC70D	$0.00
5	000308435E3E5B76	$0.00
6	000489E7EAAD463F	$0.00
7	0004F0ABD505251D	$0.00
8	00052705243EA128	$0.00
9	00070B63745BE497	$0.00
10	0007F12A492FD25D	$187,000.00

Chapter Summary

In this chapter, we used our claims and enrollment data to program measurements of payment. Specifically, we:

- Discussed the importance of measuring payment, including the fact that reducing cost without affecting quality is a hot topic and that the measurement of utilization is linked to the measurement of cost.
- Briefly discussed the concepts of risk adjustment and price standardization.
- Programmed an algorithm to measure payments for evaluation and management (E&M) services in a physician office setting by drawing on the code we wrote in Chapter 7 to define E&M services.
- Programmed an algorithm to measure payments for acute short stay hospitals using PROC SQL and the inpatient claims data set we created in Chapter 7.
- Programmed a macro algorithm to summarize total payments by beneficiary for inpatient, skilled nursing facilities, home health agencies, and hospice claims, and included conditional processing in the macro language.
- Programmed an algorithm to calculate total Part A payments for each beneficiary by aggregating total payments from each institutional claim type for the beneficiaries in our study population.

Exercises

1. Can you combine the code to identify E&M services in Chapter 7 with the code to identify E&M payments in this chapter to summarize utilization and payments in a single DATA step?
2. Can you modify the code to summarize E&M payments by provider instead of by beneficiary?
3. The output for Step 8.2 is different from that of Step 8.4. Can you explain why?
4. Can you calculate the payment by beneficiary for ambulance services?
5. How would you calculate total payments for each claim type and total Part A and Part B payments?

---

[1] These summaries are very general, and taken from excellent information provided by MedPAC, available on MedPAC's Medicare Background webpage at http://www.medpac.gov/payment_basics.cfm.

[2] Note that we could have kept these variables in Chapter 7 and simply excluded them from our analytic file because they are not summary information.

[3] For more information on the **PMT_AMT** variable, see the definition of Claim Payment Amount available on ResDAC's website at http://www.resdac.org/cms-data/variables/Claim-Payment-Amount-0.

[4] For more information on negative payment amounts, see the article called Negative Payment Amounts in the Medicare Claims Data, available on ResDAC's website at http://www.resdac.org/resconnect/articles/120.

[5] Technically, HHAs are paid under both Part A and Part B. Specifically, some services and supplies are paid under Part B, like visits after the hundredth visit, some drugs, and some DME supplies. For simplicity's sake, we are assuming that all HHA expenses are paid under Part A.

[6] Consider using options like mprint and mlogic to better understand the execution of this macro that contains conditional logic.

[7] It is important to note that size of the data sets and efficiency may become a consideration here. One possible route around the size issue is to summarize each data set separately, and then simply sum the seven total payment amounts (one for each claim type) into a total Part A payment amount. Because the input data for this step contains one record for each beneficiary (and two variables for each of these records), it should be easy to calculate how large your data set will become.

Chapter 9: Programming to Identify Chronic Conditions

Introduction and Goals

In simple terms, chronic conditions are medical conditions which have some permanence and can progressively worsen over time, thus requiring long-term management to monitor or treat (e.g., control symptoms or otherwise influence the course of the condition). Chronic conditions are not curable by any short-term treatment, and can exhibit remissions and exacerbations.[1] Commonly identified chronic conditions in the Medicare beneficiary population include Alzheimer's disease, asthma, chronic kidney disease, chronic obstructive pulmonary disease (COPD), depression, diabetes, glaucoma, and certain types of cancer, like breast cancer or colorectal cancer.[2]

Chronic conditions are important to study for several reasons. First, many Medicare beneficiaries are identified as having at least one chronic condition. For example, roughly 4 million Medicare beneficiaries are identified as having COPD and about 10 million beneficiaries are identified as diabetic.[3] Second, beneficiaries with chronic conditions require a high level of specialized and concerted medical care, often resulting in the generation of many claims. In other words, utilization of services for beneficiaries with chronic conditions will be relatively higher than average, and cost may follow suit. Finally, beneficiaries identified as having chronic conditions are commonly used in the measurement of quality outcomes because they benefit from certain standard medical practices. For example, diabetics benefit from annual dilated retinal eye exams to monitor changes in the retina (called retinopathy) that can seriously harm one's vision. Studying whether a provider performs this test on diabetic patients is a good way of measuring the quality of care provided by the doctor.

In this chapter, we will advance our discussions from Chapter 7 and Chapter 8 by providing more detail on how to measure quality outcomes. Specifically, we will combine the study of two common chronic conditions, COPD and diabetes, with the concept of measuring patient outcomes. Admittedly, quality of care and value measurement are controversial topics, especially if the measurements play a role in the assessment of provider performance or the determination of payment to providers. Indeed, patient outcomes (like hospital readmissions) often depend on things outside of providers' control, like the behavior of the patient (e.g., does the patient exercise regularly, smoke, or reliably take prescribed medications?) and the patient's general level of health when he or she began treatment. What's more, some of these factors, such as whether or not a patient exercises regularly, cannot be measured with Medicare administrative claims data.

Given the above, our goal in this chapter is not to make any determinations about quality of care; after all, we are using fake data purely for simulation purposes! Rather, our goal is to use our example research programming project to teach the reader the mechanics of how to use Medicare data to identify chronic conditions and commonly used indicators of quality outcomes. By the end of this chapter, the reader will understand important foundational concepts to use Medicare administrative claims and enrollment data to identify most any chronic condition or compute most any quality outcome metric. For example, we will discuss calculating hospital readmissions and using diagnosis and procedure codes to identify a condition or an event related to quality outcome measurement (such as diabetic eye exams). The reader could adapt the measurement of eye exams for diabetics to examine a different procedure (say, an immunization for influenza) for beneficiaries with a different chronic condition (such as heart failure). To these ends, in this chapter, we will code algorithms to do the following:

- Use our Medicare administrative claims data to identify beneficiaries with two chronic conditions, diabetes and COPD.
- Measure evaluation and management utilization for beneficiaries with diabetes or COPD.
- Measure utilization of diabetic eye exams for beneficiaries with diabetes.
- Measure hospital readmissions for beneficiaries with COPD.

As discussed in earlier chapters, the online companion to this book is at http://support.sas.com/publishing/authors/gillingham.html. Here, you will find information on creating dummy source data, the code in this and subsequent chapters, as well as answers to the exercises in this book. I expect you to visit the book's website, create your own dummy source data, and run the code yourself.

Review and Approach

Review: Peculiarities of Medicare Data

Refer to the reviews in Chapters 6, 7, and 8 to review the material relevant to this chapter's study of chronic conditions and quality. As mentioned above, quality measurement can be controversial because it sometimes involves rating provider performance. With this in mind, let's review some peculiarities and possible limitations of Medicare administrative data so we can be better prepared to interpret our output.

- Medicare coverage for the elderly is provided regardless of medical history. For example, a beneficiary diagnosed with COPD who becomes eligible for Medicare is granted coverage regardless of her COPD diagnosis. If you are used to working with commercial healthcare claims data, you will likely notice some unique characteristics of the Medicare population, like more beneficiaries identified as having at least one chronic condition (though this may change with the implementation of the Affordable Care Act).
- Some services may not appear in Medicare administrative data. For example, prescription drugs administered during a hospital stay may not appear in the claims data at all. Additionally, services paid for by Medicare Part C may not appear in the administrative claims files because managed care providers pay them.
- The administrative data we use for research purposes is updated on a regular basis, but only with claims that have been received, adjudicated, and deemed final action. As such, the files we use at any given time do not contain all final action claims submitted and paid up to the date of extraction of the data. In some rare cases, when not impractical, medical record reviews can fill in any gaps in information. What's more, we will see changes to the information available for quality measurement as electronic medical records become more prevalent.
- The correct composition of any study population must be determined carefully. In Chapter 6, we determined that our example research project will examine only those beneficiaries continuously enrolled in fee-for-service Medicare during a defined timeframe, a decision that has implications for the output of our example research programming project. More specifically, this step resulted in eliminating beneficiaries for whom we may not have a full set of claims, giving us more confidence that we are not missing claims for services like retinal eye exams or hospital readmissions for our study population.

Our Programming Plan

In Chapter 5, we submitted a data request to extract all claims and enrollment information for beneficiaries who were treated by the providers in our study population. We received and loaded our data, keeping only those variables we wished to use in our study. In Chapter 6, we used enrollment data to delimit our population to full-year FFS beneficiaries. In Chapter 7, we brought our claims data to bear on the calculation of utilization measurements. In Chapter 8, we used the Medicare amount paid to measure payments of specific services. In this chapter, we will draw on the study of utilization and payment by looking at two chronic conditions, diabetes and COPD. First, we will use diagnosis codes in specific settings of care to identify beneficiaries with diabetes or COPD (or both). Next, we will look at quality of care for these beneficiaries, giving us an excellent chance to apply the programming methods we learned in Chapters 7 and 8.

The output created in this chapter will be analytic files at the beneficiary level, which we will use for the remainder of our programming work. Looking forward, in Chapter 10, we will end our example project by summarizing all of our output.

A Note on the Chronic Conditions Data Warehouse (CCW)

Readers familiar with Medicare administrative data may be trying to reconcile two facts:

- We are coding algorithms to identify beneficiaries with two chronic conditions, diabetes and COPD, using claims data.
- CMS provides files containing this information to the research community through the Chronic Condition Data Warehouse (CCW). The CCW "is a research database designed to make Medicare, Medicaid, Assessments, and Part D Prescription Drug Event data more readily available to support research designed to improve the quality of care and reduce costs and utilization."[4] These files are available through ResDAC and CMS's data distribution contractor.

With some simplifications, we will draw on the CCW's definitions of diabetes and COPD for coding the identification of diabetes and COPD in our beneficiary population. But why do we code these algorithms ourselves when flags identifying beneficiaries with certain chronic conditions are provided by CMS? Coding chronic conditions is useful for instructional purposes, providing a context for understanding and modifying definitions of chronic conditions not available in the CCW. In addition, some investigators choose to define chronic conditions because they wish to modify the definition used in the CCW. For example, the CCW definition of diabetes includes secondary diabetes, which may not be the right choice for your particular study. It should go without saying that readers are encouraged to take full advantage of the data offered in the CCW.

Algorithms: Identifying Beneficiaries with Diabetes or COPD

Our first task is to identify beneficiaries in our population with diabetes or COPD. At the risk of being repetitive, before we begin, it is very important to note that the diagnosis codes we will use can change from year to year. The same goes for any of the codes we use in this book. For example, procedure and diagnosis codes are updated annually (new codes are created and some codes are retired). Therefore, it is good procedure to regularly check the codes you use to identify diagnoses or services (or anything, for that matter) to ensure they are current. This will be especially true after October 2014, as all codes will change when a new diagnosis coding system, ICD-10-CM, is implemented.

The algorithm to identify beneficiaries with diabetes is adapted from instructions available from the CCW, with some simplifications (e.g., for the identification of diabetes we do not require looking across two years of data).[5] Our SAS code involves searching for specific diagnosis codes (see the below algorithms for the lists of specific diagnosis codes). These codes can occur at any position on the claim (i.e., they do not have to be primary diagnoses). The setting in which these diagnosis codes occurs is important. Specifically, if we find evidence that a beneficiary was diagnosed with diabetes on only one occasion in an inpatient or skilled nursing facility setting, we will count that beneficiary as having diabetes. In these settings, trained coding experts review medical records, consult with physicians, and only record a diagnosis if the patient actually has the condition. However, we need two separate events of a diagnosis of diabetes when searching in carrier and hospital outpatient claims.[6] Why? Because the occurrence of a single diagnosis of diabetes in these settings may be what is called a "rule out" diagnosis, where the physician is ordering a lab test to check for diabetes. The second diagnosis confirms the beneficiary does have diabetes. In all cases, the diagnoses can occur any time during a two-year period (but we will search just one year of data in our example). Finally, the reader will note that we use home health claims data, but not hospice or DME data, in our identification of beneficiaries with diabetes or COPD. We use home health data even though investigators view diagnosis codes on home health claims as somewhat unreliable. We do not use hospice claims due to the nature of care provided in a hospice setting. Lastly, for obvious reasons, the DME file cannot be used to reliably identify beneficiaries with diabetes or COPD.

The algorithm to identify beneficiaries with COPD is constructed in a similar manner (and also adapted from information provided by the CCW) to the algorithm to identify beneficiaries with diabetes, but using different diagnosis codes (and the search typically occurs over a one-year period of claims data). Therefore, we will code a single algorithm to search inpatient, SNF, outpatient, and Part B carrier claims for evidence of these diagnosis codes, setting flags where appropriate. Then, we will count the flags we created, and use these counts to identify beneficiaries with diabetes or COPD.

In Step 9.1, we present an algorithm for identifying the presence of diagnosis codes used to identify diabetes or COPD in our inpatient, SNF, home health, and outpatient claims data sets. This algorithm is similar to code used in previous chapters to identify diagnosis codes in claims data. We use a macro called CLM_LOOP and array logic to create separate flags set equal to 1 in the presence of diabetes (**DIA_IP**, **DIA_SN**, **DIA_OP**, and **DIA_HH**) and COPD (**COPD_IP**, **COPD_SN**, **COPD_OP**, and **COPD_HH**) on a particular claim. Each of these claims data sets contain 25 diagnosis code variables (**ICD_DGNS_CD1-ICD_DGNS_CD25**). In addition, we must use conditional processing for our inpatient claims data set because, again, we are only searching claims for inpatient short stay and critical access hospitals with non-negative values of **PMT_AMT**. Therefore, we use a %if clause to read the inpatient claims in UTL.IP_2010_FNL_SS. We output only those records where the flag variables are equal to 1. In Step 9.3 and Step 9.4, we will use these records as definitive proof that the beneficiary has diabetes or COPD (with the exception of the flags derived from outpatient claims data).

```
/* STEP 9.1: SEARCH INPATIENT, SNF, HOME HEALTH, AND OUTPATIENT CLAIMS
FOR DIABETES AND COPD DIAGNOSIS CODES */
/* NOTE THAT THIS MACRO WILL ERROR OUT IF CALLED FOR A CLMTYP VALUE WE
DID NOT CREATE */
%macro clm_loop(clmtyp= );
data cc_&clmtyp._search(keep=bene_id dia_&clmtyp. copd_&clmtyp.);
       %if &clmtyp.=sn or &clmtyp.=op or &clmtyp.=hh %then %do;
       set utl.&clmtyp._2010_fnl;
       %end;
       %else %if &clmtyp.=ip %then %do;
       set utl.ip_2010_fnl_ss;
```

```
        %end;
        array dgnscd(25) icd_dgns_cd1-icd_dgns_cd25;
        do i=1 to 25;
               /* diabetes */
               if substrn(dgnscd(i),1,4)='3572'
               or substrn(dgnscd(i),1,5)
in('24900','24901','24910','24911','24920','24921','24930','24931',

        '24940','24941','24950','24951','24960','24961','24970','24971','
24980','24981','24990','24991',

        '25000','25001','25002','25003','25010','25011','25012','25013','
25020','25021','25022','25023',

        '25030','25031','25032','25033','25040','25041','25042','25043','
25050','25051','25052','25053',

        '25060','25061','25062','25063','25070','25071','25072','25073','
25080','25081','25082','25083',
               '25090','25091','25092','25093','36201','36202','36641')
then dia_&clmtyp.=1;
               /* copd */
               if substrn(dgnscd(i),1,3)='496'
               or substrn(dgnscd(i),1,4)
in('4910','4911','4918','4919','4920','4928','4940','4941')
               or substrn(dgnscd(i),1,5) in('49120','49121','49122') then
copd_&clmtyp.=1;
        end;
        if dia_&clmtyp.=1 or copd_&clmtyp.=1;
run;
%mend clm_loop;
%clm_loop(clmtyp=ip);
%clm_loop(clmtyp=sn);
%clm_loop(clmtyp=op);
%clm_loop(clmtyp=hh);
```

In Step 9.2, we continue the process of identifying the presence of diagnosis codes used to identify diabetes or COPD, this time using our carrier claims data set. Once again, we search each claim for diagnosis codes that indicate the presence of diabetes or COPD. When such a diagnosis code value is found in one of the carrier file's 13 line-level[7] diagnosis codes (**LINE_ICD_DGNS_CD1-LINE_ICD_DGNS_CD13**), we set the relevant flag (**DIA_CARR**, **COPD_CARR**, or both) equal to 1. We output only those records where the flag variables are equal to 1. In Step 9.3 and Step 9.4, we will use these records to build proof that the beneficiary has diabetes or COPD.

```
/* STEP 9.2: SEARCH CARRIER CLAIMS FOR DIABETES AND COPD DIAGNOSIS CODES
*/
data cc_carr_search(keep=bene_id dia_carr copd_carr);
        set utl.carr_2010_fnl;
        array lndgns(13) line_icd_dgns_cd1-line_icd_dgns_cd13;
        do i=1 to 13;
                /* diabetes */
                if substrn(lndgns(i),1,4)='3572'
                or substrn(lndgns(i),1,5)
in('24900','24901','24910','24911','24920','24921','24930','24931',

        '24940','24941','24950','24951','24960','24961','24970','24971','
24980','24981','24990','24991',

        '25000','25001','25002','25003','25010','25011','25012','25013','
25020','25021','25022','25023',

        '25030','25031','25032','25033','25040','25041','25042','25043','
25050','25051','25052','25053',

        '25060','25061','25062','25063','25070','25071','25072','25073','
25080','25081','25082','25083',
                '25090','25091','25092','25093','36201','36202','36641')
then dia_carr=1;
                /* copd */
                if substrn(lndgns(i),1,3)='496'
                or substrn(lndgns(i),1,4) in('4910 ','
4911','4918','4919','4920','4928','4940','4941')
                or substrn(lndgns(i),1,5) in('49120','49121','49122') then
copd_carr=1;
        end;
        if dia_carr=1 or copd_carr=1;
run;
```

In Step 9.3, we use a macro to loop (called CLM_LOOP2) through all claim types processed, summarizing the flags for each beneficiary by claim type, outputting one file for each relevant claim type (**CC_CNT_IP**, **CC_CNT_SN**, **CC_CNT_OP**, **CC_CNT_HH**, and **CC_CNT_CARR**). In so doing, we create summary variables called **DIA_IP_SUM**, **DIA_SN_SUM**, **DIA_OP_SUM**, **DIA_HH_SUM** and **DIA_CARR_SUM** equal to the total number of claims with the presence of a diagnosis code for diabetes in the relevant claims data set for each beneficiary. Similarly, we create summary variables called **COPD_IP_SUM**, **COPD_SN_SUM**, **COPD_OP_SUM**, **COPD_HH_SUM** and **COPD_CARR_SUM** equal to the total number of claims with the presence of a diagnosis code for COPD in the relevant claims data set for each beneficiary. Lastly, we perform nodupkey sort of the outputted files, keeping only one record per beneficiary. We now turn our attention to using these summary variables to make a final determination on whether or not a particular beneficiary has diabetes or COPD.

```
/* STEP 9.3: COUNT OCCURRENCES OF FLAGS BY SETTING */
%macro clm_loop2(clmtyp2=);
proc sql;
       create table cc_cnt_&clmtyp2. as
       select *, sum(dia_&clmtyp2.) as dia_&clmtyp2._sum,
sum(copd_&clmtyp2.) as copd_&clmtyp2._sum
       from cc_&clmtyp2._search
       group by bene_id;
quit;

/**/proc sort data=cc_cnt_&clmtyp2. nodupkey; by bene_id; run;
%mend clm_loop2;
%clm_loop2(clmtyp2=ip);
%clm_loop2(clmtyp2=sn);
%clm_loop2(clmtyp2=op);
%clm_loop2(clmtyp2=hh);
%clm_loop2(clmtyp2=carr);
```

In Step 9.4, we see the above explanation of how to count diagnosis codes for a particular claim type come to fruition. Merging the output of Step 9.3 into a single data set, the code in this step uses the summarized diabetes and COPD flags created in Step 9.3 to identify beneficiaries with diabetes or COPD as follows:

- A beneficiary is identified as having diabetes in the presence of a single relevant diagnosis found in inpatient, SNF, or home health claims data. This evidence is found where **DIA_IP_SUM**, **DIA_SN_SUM**, or **DIA_HH_SUM** is greater than or equal to one.
- A beneficiary is also identified as having diabetes in the presence of two or more relevant diagnoses found on separate outpatient or carrier claims. This evidence is found where **DIA_OP_SUM** or **DIA_CARR_SUM** is greater than or equal to two.
- A beneficiary is identified as having COPD in the presence of a single relevant diagnosis found in inpatient, SNF, or home health claims data. This evidence is found where **COPD_IP_SUM**, **COPD_SN_SUM**, or **COPD_HH_SUM** is greater than or equal to one.

- A beneficiary is also identified as having COPD in the presence of two or more relevant diagnoses found on separate outpatient or carrier claims. This evidence is found where **COPD_OP_SUM** or **COPD_CARR_SUM** is greater than or equal to two.

```
/* STEP 9.4: IDENTIFY BENEFICIARIES WITH DIABETES OR COPD USING SUMMED
FLAGS */
data cnd.dia_copd_cond(keep=bene_id dia_flag copd_flag);
       merge cc_cnt_ip cc_cnt_sn cc_cnt_op cc_cnt_hh cc_cnt_carr;
       by bene_id;
       if dia_ip_sum>=1 or dia_sn_sum>=1 or dia_hh_sum>=1 or
dia_op_sum>=2 or dia_carr_sum>=2 then dia_flag=1; else dia_flag=0;
       if copd_ip_sum>=1 or copd_sn_sum>=1 or copd_hh_sum>=1 or
copd_op_sum>=2 or copd_carr_sum>=2 then copd_flag=1; else copd_flag=0;
       if dia_flag=1 or copd_flag=1;
run;

proc sort data=cnd.dia_copd_cond nodupkey;
       by bene_id;
run;
```

Beneficiaries with diabetes are assigned a value of 1 to the **DIA_FLAG** variable, and beneficiaries with COPD are assigned a value of 1 to the **COPD_FLAG** variable. We then de-duplicate our output data set by beneficiary. We have now identified our subpopulation of beneficiaries with diabetes or COPD, contained in the outputted data set called CND.DIA_COPD_COND. Output 9.1 displays a print of the first 10 observations in the CND.DIA_COPD_COND data set.[8]

Output 9.1: Print of Beneficiaries with Diabetes or COPD

PRINT OF BENEFICIARIES WITH DIABETES OR COPD

Obs	BENE_ID	dia_flag	copd_flag
1	0004F0ABD505251D	1	0
2	00052705243EA128	1	0
3	00070B63745BE497	1	0
4	0007F12A492FD25D	1	1
5	00139C345A104F72	0	1
6	0013E139F1F37264	1	0
7	00157F1570C74E09	1	0
8	001876C5D3F3F6AA	1	0
9	0018BD6F2F493452	1	0
10	00196F0702489342	1	0

Algorithms: Evaluation and Management Visits for Beneficiaries with Diabetes or COPD

At this point in our project, we have both a list of beneficiaries in our population with diabetes or COPD, and a data set describing the E&M utilization of beneficiaries in our population. Indeed, the data set of E&M utilization called UTL.EM_UTIL, developed in Steps 7.2 through 7.4 of the code created in Chapter 7, contains a list of E&M services for the beneficiaries in our population. Our algorithm developed in Step 9.5 acknowledges that it is easier simply to merge our list of beneficiaries with COPD or diabetes with this E&M utilization data set created in Chapter 7. The output of this step is a data set called CND.EM_COND, and contains a listing of E&M services dates for beneficiaries with diabetes or COPD. To round out our work in Step 9.5, we summarize the **EM_SVC** flag for each beneficiary using a PROC FREQ, and use SAS' Output Delivery System (ODS) to create our results.

```
/* STEP 9.5: MERGE E&M UTILIZATION DATA WITH LIST OF BENEFICIARIES WITH
DIABETES OR COPD */
data cnd.em_cond;
       merge cnd.dia_copd_cond(in=a) utl.em_util;
       by bene_id;
       if a;
run;

ods html file="C:\Users\mgillingham\Desktop\SAS
Book\FINAL_DATA\ODS_OUTPUT\Gillingham_fig9_5.html"
image_dpi=300 style=GrayscalePrinter;
ods graphics on / imagefmt=png;
title "FREQ OF EM SERVICES FOR BENEFICIARIES WITH DIABETES OR COPD";
proc freq data=cnd.em_cond;
       tables em_svc*dia_flag*copd_flag;
run;
ods html close;
```

Output 9.2 shows the frequency distribution run in Step 9.5.

Output 9.2: Frequency of EM Services

FREQ OF EM SERVICES FOR BENEFICIARIES WITH DIABETES OR COPD

The FREQ Procedure

Frequency Percent Row Pct Col Pct

Table 1 of dia_flag by copd_flag			
Controlling for em_svc=1			
	copd_flag		
dia_flag	0	1	Total
0	0 0.00 0.00 0.00	2113 7.24 100.00 37.46	2113 7.24
1	23548 80.67 86.97 100.00	3528 12.09 13.03 62.54	27076 92.76
Total	23548 80.67	5641 19.33	29189 100.00

Algorithms: Diabetic Eye Exams for Beneficiaries with Diabetes

Now that we have a taste of using claims data to analyze services for beneficiaries with chronic conditions, let's search for evidence of eye exam services for the beneficiaries we identified as having diabetes. We will keep our example simple and use a small set of procedure codes to identify retinal eye exams (presented in the code below). However, it is possible to use additional codes, like ICD-9-CM surgical procedure codes. If we were calculating an official quality measure (e.g., an NCQA HEDIS measure), we would perform exclusions on our population of beneficiaries. However, for our purposes, a simple check of the carrier file will suffice; although it would be easy enough to apply this algorithm to other settings, like inpatient claims, using algorithms developed in this chapter.

In Step 9.6, we create a data set called CND.EYE_DIA_COND that contains the beneficiary identifier (**BENE_ID**) and a flag variable called **EYE_SVC** indicating whether we found evidence of an eye exam for this beneficiary in the carrier claims. First, we merge the data set created in Step 9.4 that identifies beneficiaries with diabetes or COPD (CND.DIA_COPD_COND) with the carrier claims data set (UTL.CARR_2010_FNL). Because we are interested in eye exams for beneficiaries identified as having diabetes, we subset the CND.DIA_COPD_COND data set with a 'where' clause specifying that we retain records where **DIA_FLAG** is equal to 1. From this point forward, it is a matter of executing code to search the 13 line item procedure code variables for evidence of an eye exam service (this code should look very familiar!).[9] We keep records for only those diabetic beneficiaries for whom we found evidence of an eye exam.

```
/* STEP 9.6: PULL RECORDS FROM CARRIER FILE USING CODES FOR RETINAL EYE
EXAMS */
data cnd.eye_dia_cond(keep=bene_id eye_svc);
       merge cnd.dia_copd_cond(in=a where=(dia_flag=1))
utl.carr_2010_fnl(keep=bene_id hcpcs_cd:);
       by bene_id;
       if a;
       array hcpcscd(13) hcpcs_cd1-hcpcs_cd13;
       eye_svc=0;
       do i=1 to 13;
              if substrn(hcpcscd(i),1,5)
in('92002','92004','92012','92014','92018','92019',

       '92225','92226','92230','92235','92240','92250','92260','99203','
99204','99205','99213','99214',
              '99215','99242','99243','99244','99245') then eye_svc=1;
              leave;
       end;
       if eye_svc=1;
run;
```

In Step 9.7, we de-duplicate the output of Step 9.6, overwriting the data set called CND.EYE_DIA_COND so that it now contains just one record per beneficiary.

```
/* STEP 9.7: DEDUPLICATE THE OUTPUT OF PREVIOUS STEP TO CONTAIN ONE
RECORD PER BENEFICIARY */
proc sort data=cnd.eye_dia_cond nodupkey; by bene_id; run;
```

Output 9.3 shows the first 10 observations of the output of Step 9.7.

Output 9.3: Print of Beneficiaries Who Received Diabetic Eye Exams

PRINT OF BENEFICIARIES WHO RECEIVED DIABETIC EYE EXAMS

Obs	BENE_ID	eye_svc
1	0004F0ABD505251D	1
2	00052705243EA128	1
3	00070B63745BE497	1
4	0007F12A492FD25D	1
5	0013E139F1F37264	1
6	00157F1570C74E09	1
7	001876C5D3F3F6AA	1
8	0018BD6F2F493452	1
9	00196F0702489342	1
10	001B371802E7EF58	1

Algorithms: Hospital Readmissions for Beneficiaries with COPD

Let's look for readmissions within 30 days of discharge for a hospitalization for any reason. As you advance your skills, you will be asked to program more complicated algorithms. For example, you may be asked to calculate readmission after a specific event, like a hospitalization for COPD or Acute Myocardial Infarction (AMI). In our case, we are simply looking for a readmission within 30 days of any hospitalization for our population of beneficiaries identified as having COPD, so we do not need to worry about identifying any diagnosis codes. In some cases, the reader may wish to use both prior and subsequent years of data in order to identify admissions and subsequent readmissions that occurred later in the study year, as well as early in the study year as a result of an admission in the year preceding the study year (in our case, calendar year 2010). For our simple instructional example, this will not be necessary.

First, in Step 9.8, we merge the file identifying beneficiaries with COPD created in Step 9.4 (CND.DIA_COPD_COND) with our inpatient claims data created in Chapter 7 (UTL.IP_2010_FNL_SS). We keep only those inpatient claims for beneficiaries with COPD.

```
/* STEP 9.8: JOIN INPATIENT CLAIMS WITH LIST OF BENEFICIARIES WITH COPD
*/
data copd_ip;
       merge cnd.dia_copd_cond(in=a where=(copd_flag=1))
utl.ip_2010_fnl_ss;
       by bene_id;
       if a;
run;
```

Next in Step 9.9, we present code to look through the inpatient claims to identify admissions to a hospital that occurred within 30 days of discharge for a previous admission.[10] Specifically, we use PROC SQL code to create a data set called CND.READM_COPD_COND that contains the beneficiary identifier (**BENE_ID**), discharge date (**DISCHRGDT**), and admission date (**ADMSN_DT**) on claims for beneficiaries with COPD representing readmissions to a hospital that occur 30 days after an initial admission. Note that we only select the **BENE_ID** from the data set with the alias "a" in order to avoid a warning in our log that the variable **BENE_ID** already exists on the output data set CND.READM_COPD_COND.

```
/* STEP 9.9: IDENTIFY READMISSIONS WITHIN 30 DAYS OF ANY DISCHARGE */
proc sql;
       create table cnd.readm_copd_cond as
       select a.*, b.readm
       from copd_ip(keep=bene_id dschrgdt) a, copd_ip(keep=bene_id
admsn_dt rename=(admsn_dt=readm)) b
       where a.bene_id=b.bene_id
       having ((readm gt dschrgdt+1) and (readm-dschrgdt+1<=30));
quit;
```

Output 9.4 shows the first 10 observations of the output of Step 9.9.

Output 9.4: Print of Hospital Readmissions for Beneficiaries with COPD

PRINT OF HOSPITAL READMISSIONS FOR BENEFICIARIES WITH COPD

Obs	BENE_ID	DSCHRGDT	readm
1	0007F12A492FD25D	20100612	20100616
2	007928DE5B0C4AA5	20100107	20100114
3	00F74CF706A0BE96	20100415	20100514
4	02A663B3BA049E9D	20100801	20100830
5	0388381BFCB2C4A3	20100122	20100124
6	05039EEC14848078	20100525	20100615
7	05039EEC14848078	20100619	20100707
8	09C052549BCFC935	20100208	20100213
9	1448403EBDF7E6B5	20100101	20100129
10	154379B03FD0C14B	20100614	20100620

Chapter Summary

In this chapter, we used our claims and enrollment data to study chronic conditions and quality outcomes. Specifically, we:

- Discussed the importance of studying chronic conditions. Many Medicare beneficiaries have chronic conditions; treating chronic conditions requires a high level of specialized and concerted medical care, and the treatment of chronic conditions is often looked at from the perspective of measuring quality outcomes.
- Programmed algorithms to identify beneficiaries with diabetes or COPD by defining the occurrence of certain diagnosis codes that identify these chronic conditions (including "rule out" diagnoses), searching the line items on claims for these diagnosis codes, and counting the occurrence of these codes by claim type.
- Programmed algorithms to identify evaluation and management services for beneficiaries with diabetes or COPD by merging beneficiaries with these conditions with the files we created in Chapter 7 to define E&M visits.

- Programmed simple algorithms to identify the occurrence of diabetic eye exams by searching for procedure codes that identify retinal eye exams in the Part B carrier data for beneficiaries identified as diabetic.
- Programmed simple algorithms to identify hospital readmissions for beneficiaries with COPD within 30 days of a discharge for any hospitalization.

Exercises

1. When using carrier claims to identify beneficiaries with diabetes, investigators often look for two claims separated by one or more days. In this case, we are looking for a diabetes diagnosis on two separate claims, but those claims could occur during the same time period. Can you modify Step 9.1 through Step 9.4 to identify and count diagnoses of diabetes on carrier claims occurring on separate days?
1. Can you modify the algorithms in Step 9.1 through Step 9.4 to identify beneficiaries with prostate cancer? Define beneficiaries with prostate cancer using the specifications from the CCW Medicare Administrative Data User Guide, available for download at http://www.ccwdata.org/cs/groups/public/documents/document/ccw_userguide.pdf.
2. Can you adapt the algorithms in Step 9.8 and Step 9.9 to look for readmissions within 30, 60, and 90 days of discharge for a hospitalization for AMI? Use the inpatient claims for our population of beneficiaries with COPD as your starting point.
3. Can you use the MBSF data received from CMS's data distribution contractor (i.e., our original enrollment data, not reduced to contain information only for those beneficiaries identified as continuously enrolled in Medicare FFS for all twelve months of calendar year 2010) to identify beneficiaries that died during the year, and subsequently examine their claims data to find evidence of diabetes or COPD? Compare the mortality rates of beneficiaries with diabetes with beneficiaries with COPD, as well as the overall population of all beneficiaries in the MBSF data.

---

[1] This definition is taken from the Missouri Department of Health and Senior Services, available at http://health.mo.gov/living/families/schoolhealth/glossary.php.

[2] For a full list of the chronic conditions identified in CMS's Chronic Condition Warehouse, see the Chronic Condition Reference List, available for download from the CCW website at http://www.ccwdata.org/cs/groups/public/documents/document/ccw_conditionreferencelist2011.pdf.

[3] For this and more information on beneficiary counts by chronic condition algorithm, see the Comparison of Medicare Beneficiary Counts for Chronic Condition Algorithms, available on the CCW website at http://www.ccwdata.org/cs/groups/public/documents/document/ccw_web_table_b5.pdf.

[4] This description of the CCW is taken from the CCW website, available at http://www.ccwdata.org/index.htm.

[5] For instruction on how to identify diabetes, COPD, and other chronic conditions, see The Chronic Condition Data Warehouse Medicare Administrative Data User Guide, available for download from the CCW website at http://www.ccwdata.org/cs/groups/public/documents/document/ccw_userguide.pdf.

---

[6] When identifying beneficiaries with diabetes using carrier claims, investigators often look for two claims separated by one or more days. In this case, we are looking for a diabetes diagnosis on two separate claims, but those claims could occur during the same time period. See the exercises section for a question on identifying carrier claims occurring on separate days.

[7] Recall that the carrier file has 13 line items for each claim. In earlier chapters, we searched the 13 line-level procedure codes in carrier claims.

[8] In Chapter 10, we will examine the overall rates of diabetes and COPD in our study population, as well as rates for the study population by state and county.

[9] We use procedure codes to identify routine eye exams, but there can be more detailed definitions depending on the purpose of the investigation.

[10] This algorithm is adapted from a discussion on The University of Georgia's listserv, available at http://listserv.uga.edu/cgi-bin/wa?A2=ind0612b&L=sas-l&P=21025. It is simple, clean, and perfect for our purposes. As you perform more complicated analyses (like looking for admissions for specific reasons), you will write more complex readmission algorithms.

Chapter 10: Presenting Output and Project Disposition

Introduction

It has been a long a journey! We learned about the Medicare program, Medicare data, and knocked out all but the final steps in our example research programming project. The purpose of this chapter is to take the final steps in our research programming project by synthesizing our results and completing the project disposition phase of our project plan. As discussed in earlier chapters, the online companion to this book is at http://support.sas.com/publishing/authors/gillingham.html. Here, you will find information on creating dummy source data, the code in this and subsequent chapters, as well as answers to the exercises in this book. I expect you to visit the book's website, create your own dummy source data, and run the code yourself.

At this point in our project, it is easy to think we are nearly finished! After all, we performed some serious data manipulation to create analytic files for enrollment, utilization, and cost analyses, including looking at chronic conditions and quality measurement. We positioned our project well; all we need to do is deliver some output, right? Well, not so fast! By now, you probably realize that nothing is as easy as it first appears! As investigators, we need to put a lot of thought into presenting the information stored in our analytic files in a meaningful way. Then, after presenting our results, we must close out the project. The process of ending a project methodically is called project disposition; disposition requires ample time and effort to ensure that our specifications, code, and results will be clear to us for years to come. In this way, the proper disposition of our example research programming project brings an extra layer of legitimacy to our work, and facilitates our ability to use our work in the future.

Synthesis – Review and Approach

Review: Our Accomplishments

Before we begin, instead of our usual review of topics relevant to our programming work, we will reward ourselves by reviewing all of our fantastic accomplishments! Recall that we have undertaken an effort to evaluate the effectiveness of a pilot program that is designed to incentivize providers to reduce costs to Medicare and improve quality outcomes. Our evaluation effort hinges on measuring utilization, cost to Medicare, and quality by using a list of beneficiaries who interacted with the providers in our sample population; recall that this list of providers and beneficiaries was provided to us at the outset of our study. To these ends, we performed the following work:

- In Chapters 4 and 5, we established our project plan. In addition, we requested, received, loaded, and transformed the data used in subsequent chapters.
- In Chapter 6, we created a file that contained demographic information for beneficiaries who were continuously enrolled in Medicare FFS throughout all twelve months of calendar year 2010.
- In Chapter 7, we created analytic files that contained summaries of utilization of services for our population of continuously enrolled beneficiaries.
- In Chapter 8, we created analytic files that contained summaries of the costs to Medicare for services for our population of continuously enrolled beneficiaries.
- In Chapter 9, we identified beneficiaries with diabetes or COPD. In addition, we created analytic files that focused on simple measurements of quality outcomes.

Our Programming Plan

Until now, our programming adventure has focused on calculating the information needed to answer the research questions we posed at the beginning of our example project. Now, we must slightly shift our focus to finish our work by presenting our utilization, cost, and quality outcome measurements. To this end, we will utilize data from some of the analytic files that we created in prior chapters, and use some of the work that we performed in Chapter 6 to add state and county names to our file of continuously enrolled beneficiaries.

The algorithms that we will develop in the chapter are by no means exhaustive, and we do not endeavor to summarize all of the analyses that we computed in prior chapters. We could think of many, many additional ways of looking at the output we created. For example, we could present analyses of outcomes that are broken out by provider specialty; we could compare the measurements for the population of providers in our finder file against other providers that did not participate in the program we are studying; or, we could compare results over time to look for significant changes in outcomes. The purpose of this text is to lay the groundwork for the reader to be able to independently answer these questions. Therefore, one of the chapter exercises listed below asks you to think of some additional questions we might answer with our data.

Algorithms: Presenting Selected Measurements of Utilization, Cost to Medicare, and Quality Outcomes

Let's start by simply displaying the following pieces of our utilization, cost to Medicare, and quality outcomes of the work that we performed in Chapter 7 through Chapter 9, respectively:

- Drawing on our work in Chapter 7, we will present ambulance utilization information by beneficiary age.
- Drawing on our work in Chapter 8, we will present Medicare payments for E&M services summarized by the providers in our study population.
- Drawing on our work in Chapter 9, we will present rates of diabetes and COPD in our population of continuously enrolled beneficiaries.

Again, we could have easily chosen to present other information that we calculated in previous chapters, like the total costs to Medicare that were associated with home health agency services. One of the chapter exercises listed below asks you to choose and present information that was calculated in Chapter 7, Chapter 8, or Chapter 9 other than what was chosen as an example for our work in this section.

Presenting Ambulance Utilization

One interesting presentation of output is to examine beneficiaries with at least one claim for ambulance services, along with the age of each beneficiary. This type of algorithm provides a good example for analyzing relevant summary information that may be used to guide further analyses.

In Step 10.1, we begin our work to present ambulance utilization information by merging our file of continuously enrolled beneficiaries (ENR.CONTENR_2010_FNL) with the analytic summary file UTL.AMB_UTIL created in Step 7.11 of Chapter 7 (we also bring in our formatting of age categories that you will recognize from our work in Chapter 6). The file UTL.AMB_UTIL contains one record per beneficiary with at least one claim for an ambulance service. Because both files are already sorted by **BENE_ID**, there is no need for a pre-merge sort. The outputted file, called FNL.PRESENT_AMB, contains one record for each of the beneficiaries in UTL.AMB_UTIL (we use an "if b" merge), as well as the **STUDY_AGE** and **AGE_CATS** variables (kept from the ENR.CONTENR_2010_FNL file) and the **TOT_AMB_SVC** variable (kept from the ambulance utilization analytic file).

```
/* STEP 10.1: MERGE CONTINUOUSLY ENROLLED BENEFICIARIES WITH AMBULANCE
UTILIZATION DATA */
proc format;
    value age_cats_fmt
        0='AGE LESS THAN 65'
        1='AGE BETWEEN 65 AND 74, INCLUSIVE'
        2='AGE BETWEEN 75 AND 84, INCLUSIVE'
        3='AGE BETWEEN 85 AND 94, INCLUSIVE'
        4='AGE GREATER THAN OR EQUAL TO 95';
run;

data fnl.present_amb(keep=bene_id tot_amb_svc study_age age_cats);
    merge enr.contenr_2010_fnl(in=a) utl.amb_util(in=b);
       by bene_id;
       if b;
run;
```

Next, in Step 10.2, we use ODS to present a PROC SUMMARY of the beneficiaries with at least one claim for an ambulance service (using a 'where' clause that the value of **TOT_AMB_SVC** is not null), along with the number of ambulance services (the value of the **TOT_AMB_SVC** variable) and the beneficiary's age category (the value of **AGE_CATS**). Note that we must first sort the file by **AGE_CATS**, and we utilize the formatting of the **AGE_CATS** variable in our presentation of the output.

```
/* STEP 10.2: PRESENT MEASUREMENT OF AMBULANCE UTILIZATION */
proc sort data=fnl.present_amb;
       by age_cats;
run;

ods html file="C:\Users\mgillingham\Desktop\SAS
Book\FINAL_DATA\ODS_OUTPUT\Gillingham_fig10_2.html"
image_dpi=300 style=GrayscalePrinter;
ods graphics on / imagefmt=png;
title "LISTING OF BENEFICIARIES WITH AMBULANCE UTILIZATION GROUPED BY
AGE_CATS";
proc summary data=fnl.present_amb(where=(tot_amb_svc ne .)) print;
       var tot_amb_svc;
       by age_cats;
    format age_cats age_cats_fmt.;
run;
ods html close;
```

Output 10.1 is a snapshot of some of the output of Step 10.2.

Output 10.1: List of Beneficiaries with Ambulance Utilization Grouped by AGE_CATS

LISTING OF BENEFICIARIES WITH AMBULANCE UTILIZATION GROUPED BY AGE_CATS

The SUMMARY Procedure

Beneficiary age category at beginning of reference year (January 1, 2010)=AGE LESS THAN 65

Analysis Variable : tot_amb_svc				
N	Mean	Std Dev	Minimum	Maximum
2615	3.6493308	9.4003743	1.0000000	221.0000000

Beneficiary age category at beginning of reference year (January 1, 2010)=AGE BETWEEN 65 AND 74, INCLUSIVE

Analysis Variable : tot_amb_svc				
N	Mean	Std Dev	Minimum	Maximum
5878	3.4443688	8.5163597	1.0000000	292.0000000

Beneficiary age category at beginning of reference year (January 1, 2010)=AGE BETWEEN 75 AND 84, INCLUSIVE

Analysis Variable : tot_amb_svc				
N	Mean	Std Dev	Minimum	Maximum
5275	3.4684360	8.0200438	1.0000000	316.0000000

Beneficiary age category at beginning of reference year (January 1, 2010)=AGE BETWEEN 85 AND 94, INCLUSIVE

Analysis Variable : tot_amb_svc				
N	Mean	Std Dev	Minimum	Maximum
2431	3.4006582	5.8517965	1.0000000	88.0000000

Beneficiary age category at beginning of reference year (January 1, 2010)=AGE GREATER THAN OR EQUAL TO 95

Analysis Variable : tot_amb_svc				
N	Mean	Std Dev	Minimum	Maximum
608	4.0641447	10.7670021	1.0000000	214.0000000

Presenting Medicare Payments for Evaluation and Management Services by Provider

Another interesting presentation of the summary analytic files created in previous chapters is a listing of summary statistics related to costs to Medicare for evaluation and management visits. In this example, we use the file that contains providers and their associated beneficiaries (described in Chapter 5 as our "finder file" or "attribution file") in our analysis. Recall we received this file at the beginning of our example research programming project, and we used it to guide our project's data extraction process. Now, we use this attribution file to examine costs for the sample of providers in our study.

We measured total E&M costs to Medicare using the code presented in Step 8.2 of Chapter 8, outputting a data set called CST.EM_CST_BENE that contains one record for each beneficiary, along with a variable called **EM_TOTCOST_BENE** that is a summary of each beneficiary's total E&M costs. Now, in Step 10.3, we merge the CST.EM_CST_BENE data set with our attribution file (called SRC.FINDER_ATTRIB) in order to "attribute" a provider to each beneficiary in the E&M cost summary data set. Both data sets are already sorted by **BENE_ID**, so pre-merge sorting is not necessary. We keep only those records in the CST.EM_CST_BENE data set (i.e., we perform an "if b" merge) because those are the beneficiaries for whom we have E&M cost information. We output a data set called PRESENT_EMCOST that contains the E&M cost information (**EM_TOTCOST_BENE**) for each continuously enrolled beneficiary with at least one E&M service, as well as the identifier variable of the provider associated with that beneficiary (**PRFNPI**).

```
/* STEP 10.3: MERGE E&M COST INFORMATION WITH PROVIDER ATTRIB FILE */
data present_emcost(keep=bene_id prfnpi em_totcost_bene);
    merge src.finder_attrib(in=a) cst.em_cost_bene(in=b);
    by bene_id;
    if b then output present_emcost;
run;
```

Finally, in Step 10.4, we use PROC SQL to summarize the E&M cost information (**EM_TOTCOST_BENE**) contained in the FNL.PRESENT_EMCOST data set by provider (**PRFNPI**).We call our summarized cost **EM_TOTCOST_PROV**. We also summarize the number of relevant beneficiaries into a variable called **EM_TOTBENE_PROV**. We output a file called FNL.PRESENT_EMCOST.

```
/* STEP 10.4: PRESENT MEASUREMENT OF E&M COSTS TO MEDICARE BY PROVIDER
*/
proc sql;
       create table fnl.present_emcost as
       select prfnpi, count(distinct bene_id) as em_totbene_prov,
sum(em_totcost_bene) as
              em_totcost_prov format=dollar15.2 label='TOTAL E&M COST FOR
PROVIDER'
       from present_emcost
       group by prfnpi;
quit;
```

Output 10.2 is a snapshot of some of the output of Step 10.4, displayed in descending order by **EM_TOTBENE_PROV**.

Output 10.2: PROC PRINT of EM Costs to Medicare, Summed by Provider

PROC PRINT OF EM COSTS TO MEDICARE, SUMMED BY PROVIDER

Obs	PRFNPI	em_totbene_prov	em_totcost_prov
2	0888901330	344	$97,040.00
3	8207899456	241	$59,120.00
4	9979265126	227	$58,990.00
5	2052541087	184	$45,060.00
6	1973668724	178	$47,070.00
7	7961400834	154	$38,060.00
8	3552796390	148	$35,730.00
9	1052105939	143	$36,730.00
10	2866295111	143	$34,410.00
11	0329757128	137	$35,650.00

Presenting Rates of Diabetes and COPD

Finally, let's utilize our work that identified beneficiaries with diabetes or COPD by studying rates of these conditions in our population of continuously enrolled beneficiaries. This exercise provides a good example of presenting results that provide a framework for additional analyses. For example, if there are a small number of beneficiaries with diabetes in our overall population, the population of a particular county or state, or a population of beneficiaries attributed to a certain provider, we could surmise that some utilization and cost measurements may follow suit.

In Step 10.5, we begin our look at rates of diabetes and COPD in our population of continuously enrolled beneficiaries by merging our file containing a list of beneficiaries with COPD or diabetes with our population of continuously enrolled beneficiaries. In Step 9.1 through Step 9.4 presented in Chapter 9, we created a file called CND.DIA_COPD_COND. This file contains one record per beneficiary with either diabetes or COPD (or both diabetes and COPD). Because this data set contains records only for beneficiaries with diabetes or COPD, we must merge the file with our population of continuously enrolled beneficiaries to learn about the number of beneficiaries with diabetes or COPD in our population as a whole. To this end, we merge the CND.DIA_COPD_COND data set with the same ENR.CONTENR_2010_FNL data set used above, keeping all records in the ENR.CONTENR_2010_FNL data set to create the data set FNL.PRESENT_DIACOPD. The values of the variables identifying beneficiaries with diabetes or COPD in the data set CND.DIA_COPD_COND are set equal to zero for beneficiaries in ENR.CONTENR_2010_FNL who are not in CND.DIA_COPD_COND (indicating that these beneficiaries were not identified as having diabetes or COPD).

```
/* STEP 10.5: MERGE CONTINUOUSLY ENROLLED BENEFICIARIES WITH POPULATION
OF DIABETICS AND BENES WITH COPD */
data fnl.present_diacopd(keep=bene_id dia_flag copd_flag);
    merge enr.contenr_2010_fnl(in=a) cnd.dia_copd_cond(in=b);
    by bene_id;
       if a;
       if dia_flag=. then dia_flag=0; if copd_flag=. then copd_flag=0;
run;
```

In Step 10.6, we present a simple frequency distribution of beneficiaries with diabetes or COPD (or both diabetes and COPD) using a PROC FREQ of the data set FNL.PRESENT_DIACOPD that we created in Step 10.5. We perform frequency distribution on the variable that identifies beneficiaries with diabetes (**DIA_FLAG**) and the variable that identifies beneficiaries with COPD (**COPD_FLAG**). In addition, we perform a cross-tabulation of these variables. As in previous examples in this chapter, we use ODS to output the PROC FREQ results to the same folder where our permanent data is stored for our work in this chapter.

```
/* 10.6: PRESENT RATES OF DIABETES, COPD, AND BOTH DIABETES AND COPD */
ods html file="C:\Users\mgillingham\Desktop\SAS
Book\FINAL_DATA\ODS_OUTPUT\Gillingham_fig10_6.html"
image_dpi=300 style=GrayscalePrinter;
ods graphics on / imagefmt=png;
title "RATES OF DIABETES AND COPD IN CONTINUOUSLY ENROLLED BENEFICIARY
POPULATION";
proc freq data=fnl.present_diacopd;
    tables dia_flag copd_flag / missing;
run;
ods html close;
```

Output 10.3 shows the results of Step 10.6.

Output 10.3: Rates of Diabetes and COPD in Continuously Enrolled Beneficiary Population

RATES OF DIABETES AND COPD IN CONTINUOUSLY ENROLLED BENEFICIARY POPULATION

The FREQ Procedure

dia_flag	Frequency	Percent	Cumulative Frequency	Cumulative Percent
0	38315	57.02	38315	57.02
1	28883	42.98	67198	100.00

copd_flag	Frequency	Percent	Cumulative Frequency	Cumulative Percent
0	61228	91.12	61228	91.12
1	5970	8.88	67198	100.00

Algorithms: Presenting Inpatient Length of Stay Information by State

In this section, we build on our work presenting the work of previous chapters by displaying the inpatient length of stay by state. To this end, we use the data set UTL.STAY_UTL that we created in Step 7.5 through Step 7.8 presented in Chapter 7. This data set contains summary information on length of stay, but it also contains one record for each beneficiary's unique inpatient claim admission and discharge dates (more on this below). We also use the data set of continuously enrolled beneficiaries used in the examples above. This data set, called ENR.CONTENR_2010_FNL, contains information on each beneficiary's state of residence (added to the ENR.CONTENR_2010_FNL data set in Chapter 6).

As mentioned above, the UTL.STAY_UTL data set contains one record for each beneficiary's unique inpatient claim admission and discharge dates. We do not need to use the admission and discharge date information for our analysis. Therefore, in Step 10.7, we sort the UTL.STAY_UTIL data set, removing duplicate records for each beneficiary, and keeping only the beneficiary identifier and summary information pertaining to each beneficiary's hospital stays. The output, a data set called STAY_UTIL, contains one record for each beneficiary, as well as each beneficiary's inpatient length of stay summary information.

```
/* STEP 10.7: REMOVE STAY DATES AND SORT IP STAY UTILIZATION DATA */
proc sort data=utl.stay_util(keep=bene_id stay_cnt tot_ip_los ip_alos)
out=stay_util nodupkey;
      by bene_id;
run;
```

In Step 10.8, we add the name of the beneficiary's state to the STAY_UTIL data by merging it with the ENR.CONTENR_2010_FNL data set. We are presenting a summary of length of stay only for those beneficiaries with at least one inpatient claim, so we do not retain all beneficiaries in the ENR.CONTENR_2010_FNL data set (i.e., we use an "if b" merge to ensure we retain only the beneficiaries with at least one hospital stay). We call the output of this merge FNL.PRESENT_IPLOS.

```
/* 10.8: MERGE INPATIENT LENGTH OF STAY DATA WITH BENEFICIARY SSA STATE
INFORMATION */
data fnl.present_iplos(keep=bene_id stay_cnt tot_ip_los ip_alos state);
    merge enr.contenr_2010_fnl(in=a) stay_util(in=b);
    by bene_id;
    if b;
run;
```

In Step 10.9, we sort the FNL.PRESENT_IPLOS data set in preparation for displaying data by state. Specifically, we sort the FNL.PRESENT_IPLOS by the variable **STATE**, which will be the 'by' variable in the PROC FREQ performed in the next step.

```
/* STEP 10.9: SORT DATA BY SSA STATE IN PREPARATION FOR PRESENTATION OF
IP LOS BY SSA STATE AND COUNTY */
proc sort data=fnl.present_iplos;
       by state;
run;
```

Finally, in Step 10.10, we perform a PROC FREQ to output measurement of the average inpatient length of stay for each beneficiary (**IP_LOS**), by the name of the state (**STATE**) associated with the beneficiary's SSA state code provided in the MBSF data. Once again, we use ODS to output the PROC FREQ results to the same folder where our permanent data is stored for our work in this chapter. Note that we limit our frequency to the state of Michigan, and to values of **IP_ALOS** that are between 1 and 5 days.

```
/* STEP 10.10: PRESENT IP LOS INFORMATION BY BENEFICIARY SSA STATE */
ods html file="C:\Users\mgillingham\Desktop\SAS
Book\FINAL_DATA\ODS_OUTPUT\Gillingham_fig10_10.html"
image_dpi=300 style=GrayscalePrinter;
ods graphics on / imagefmt=png;
title "FREQUENCY OF IP LOS BY SSA STATE";
proc freq data=fnl.present_iplos(where=(state='MI' and 1<=ip_alos<=5));
    tables ip_alos;
    by state;
run;
ods html close;
```

Output 10.4 shows the output of Step 10.10.

Output 10.4: Frequency of IP LOS by SSA State

FREQUENCY OF IP LOS BY SSA STATE

The FREQ Procedure

state=MI

ip_alos	Frequency	Percent	Cumulative Frequency	Cumulative Percent
1	4	2.84	4	2.84
2	33	23.40	37	26.24
3	34	24.11	71	50.35
3.3333333333	1	0.71	72	51.06
3.5	1	0.71	73	51.77
4	45	31.91	118	83.69
4.5	3	2.13	121	85.82
5	20	14.18	141	100.00

Algorithms: Presenting Mean Medicare Part A Payments per Beneficiary by State and County

Understanding the landscape of costs to Medicare adds an important dimension to our analyses. For example, if our population contains many beneficiaries identified as having COPD, we might expect payments made by Medicare to be higher for this study population. In addition, understanding total costs to Medicare for the beneficiaries in our study population by geographic region opens the gateway to further analyses. For example, we could combine some of the techniques developed in these steps and apply them to a study of total costs to Medicare by chronic condition, or by provider in our study population.

In Step 10.11, we add beneficiary state and county information to our data set of total Part A costs to Medicare. This data set containing summary information on costs, called CST.PA_TOTCOST_BENE, was created in Chapter 8. Our merge to add state and county information to this data set looks the same as the merge presented in Step 10.8, except that we retain all records in the ENR.CONTENR_2010_FNL data set (i.e., we perform an "if a" merge). In addition, we add a line of code to set the value of the variable summarizing total Part A costs to Medicare (**PA_TOTCOST_BENE**) equal to 0 for beneficiaries without a record in the ENR.CONTENR_2010_FNL data set because it is possible (though unlikely) for a beneficiary to have an enrollment record without any corresponding claims. The output of this merge, a data set called FNL.PRESENT_PACOST, contains one record for each continuously enrolled beneficiary, along with each beneficiary's total Part A cost summary information, and information on the beneficiary's state and county of residence.

```
/* 10.11: MERGE TOTAL PART A COST DATA WITH BENEFICIARY SSA STATE AND
COUNTY INFORMATION */
data fnl.present_pacost(keep=bene_id pa_totcost_bene state county);
    merge enr.contenr_2010_fnl(in=a) cst.pa_totcost_bene(in=b);
    by bene_id;
    if a;
       if a and not b then pa_totcost_bene=0;
run;
```

In Step 10.12, we sort the output of Step 10.11 by **STATE** and **COUNTY** in preparation for displaying total Part A costs to Medicare by **STATE** and **COUNTY** in the PROC MEANS presented in the next step.

```
/* STEP 10.12: SORT DATA BY SSA STATE AND COUNTY IN PREPARATION FOR
PRESENTATION OF COST BY SSA STATE AND COUNTY */
proc sort data=fnl.present_pacost;
       by state county;
run;
```

Finally, in Step 10.13, we perform a PROC MEANS on our summaries of total Part A costs to Medicare in order to present mean Medicare payments per beneficiary by state and county name. As in the prior example, we use ODS to output the results to the same folder where our permanent data is stored for our work in this chapter, and we limit the output to those beneficiaries who reside in a single county in Michigan. We also ask for the minimum and maximum values for each state and county combination.

```
/* STEP 10.13: PRESENT TOTAL PART A COST BY BENEFICIARY SSA STATE AND
COUNTY */
ods html file="C:\Users\mgillingham\Desktop\SAS
Book\FINAL_DATA\ODS_OUTPUT\Gillingham_fig10_13.html"
image_dpi=300 style=GrayscalePrinter;
ods graphics on / imagefmt=png;
title "PROC MEANS OF TOTAL PART A COST TO MEDICARE BY SSA STATE AND
COUNTY";
proc means data=fnl.present_pacost(where=(state='MI' and
county='WASHTENAW'))
```

```
                n nmiss mean min max maxdec=1;
     var pa_totcost_bene;
     by state county;
run;
ods html close;
```

Output 10.5 shows the results of Step 10.13, delimited to print for one county in Michigan.

Output 10.5: PROC MEANS of Total Part A Cost to Medicare by SSA State and County

PROC MEANS OF TOTAL PART A COST TO MEDICARE BY SSA STATE AND COUNTY

The MEANS Procedure

state=MI county=WASHTENAW

Analysis Variable : pa_totcost_bene TOTAL PART A COSTS PER BENEFICIARY				
N	N Miss	Mean	Minimum	Maximum
65	0	4123.1	0.0	72000.0

Algorithms: Presenting Rates of Diabetic Eye Exams

In our final example, we present the occurrence of diabetic eye exams by SSA state and county name. We wish to limit our presentation only to those continuously enrolled beneficiaries who are identified as diabetic. This work incorporates the work of previous chapters to identify beneficiaries who have diabetes, and then requires us identify diabetic eye exam services for these diabetic beneficiaries.

Specifically, we utilize the data set CND.EYE_DIA_COND that we created in Chapter 9 and use the variable **EYE_SVC,** which flags evidence that a diabetic beneficiary has had a service for a diabetic eye exam. However, the data set CND.EYE_DIA_COND contains records only for beneficiaries identified as having a diabetic eye exam. Because we want to look at the occurrence of diabetic eye exams in the overall population of continuously enrolled beneficiaries identified as diabetic, we must utilize the data set CND.DIA_COPD_COND that contains a list of beneficiaries identified as having diabetes. Finally, we utilize the ENR.CONTENR_2010_FNL data set to add beneficiary state and county name information to our analysis.

To this end, in Step 10.14, we perform a merge as follows:

- We merge three data sets (CND.EYE_DIA_COND, CND.DIA_COPD_COND, and ENR.CONTENR_2010_FNL).

- Using the data set CND.DIA_COPD_COND, we set the population of our output file to be only those continuously enrolled beneficiaries who were identified as having diabetes.
- We use the data set CND.EYE_DIA_COND to add the EYE_SVC flag to the output data set.
- We use the data set ENR.CONTENR_2010_FNL to add HRR information for each beneficiary in the output data set.
- We output a data set called FNL.PRESENT_EYEDIA, keeping the state and county name variables (**STATE** and **COUNTY**), as well as the **EYE_SVC** variable. Note that the missing values of EYE_SVC (created for those continuously enrolled beneficiaries with diabetes who did not receive a diabetic eye exam) are set equal to zero.

```
/* STEP 10.14: MERGE CONTINUOUS ENROLLMENT AND DIABETIC EYE EXAM
INFORMATION */
data fnl.present_eyedia(keep=bene_id eye_svc state county);
    merge enr.contenr_2010_fnl(in=a) cnd.eye_dia_cond(in=b)
cnd.dia_copd_cond(in=c keep=bene_id dia_flag);
       by bene_id;
    if c and dia_flag=1;
       if eye_svc=. then eye_svc=0;
run;
```

In Step 10.15, we sort the output of Step 10.14 by **STATE** and **COUNTY** in preparation for displaying the number of diabetic eye exams by **STATE** and **COUNTY** in the PROC FREQ presented in the next step.

```
/* STEP 10.15: SORT DATA BY SSA STATE AND COUNTY IN PREPARATION FOR
PRESENTATION BY SSA STATE AND COUNTY */
proc sort data=fnl.present_eyedia;
       by state county;
run;
```

Finally, in Step 10.16, we perform two separate PROC FREQs on the **EYE_SVC** variable. First, we create a frequency distribution of the rates of diabetic eye exams in the overall population of continuously enrolled beneficiaries with diabetes. Next, we perform the same frequency distribution by **STATE** and **COUNTY**, limiting our analysis to a single county in Michigan. You can see that there are few beneficiaries in the overall population who did not receive a diabetic eye exam, and there are no beneficiaries in Washtenaw County Michigan who did not receive a diabetic eye exam. We use ODS to output the results to the same folder where our permanent data is stored for our work in this chapter.

```
/* STEP 10.16: PRESENT MEASUREMENT OF DIABETIC EYE EXAM SERVICES BY
OVERALL AND BY SSA STATE AND COUNTY */
ods html file="C:\Users\mgillingham\Desktop\SAS
Book\FINAL_DATA\ODS_OUTPUT\Gillingham_fig10_16.html"
image_dpi=300 style=GrayscalePrinter;
ods graphics on / imagefmt=png;
title "RATES OF DIABETIC EYE EXAMS";
proc freq data=fnl.present_eyedia;
```

```
    tables eye_svc / missing;
run;

proc freq data=fnl.present_eyedia(where=(state='MI' and
county='WASHTENAW'));
    tables eye_svc / missing;
    by state county;
run;
ods html close;
```

Output 10.6 shows the results of Step 10.16.

Output 10.6: Rates of Diabetic Eye Exams

RATES OF DIABETIC EYE EXAMS

The FREQ Procedure

eye_svc	Frequency	Percent	Cumulative Frequency	Cumulative Percent
0	2168	7.51	2168	7.51
1	26715	92.49	28883	100.00

RATES OF DIABETIC EYE EXAMS

The FREQ Procedure

state=MI county=WASHTENAW

eye_svc	Frequency	Percent	Cumulative Frequency	Cumulative Percent
1	22	100.00	22	100.00

Production Execution in Batch

As stated below, all programs should be executed in batch mode such that the logs and lists have the same name as their parent program (but a different file extension, of course). In addition, it is helpful to undertake a "production" execution of the code that entails running each program in sequence without breaks so that the date and time stamps on the files are sequential. Depending on your operating system, executing code in batch can be as simple as "right clicking" on the file you wish to execute and selecting the option to execute in batch. Upon completion, log and list files are generated automatically. The reader should carefully review the logs for errors, warnings, and uninitialized variables (this can be accomplished quite easily by performing a find on "error," "warning," and "uninitialized". If the reader is operating in a UNIX environment, he or she can execute a script in the UNIX shell to execute the code in batch. Prior to executing in batch, I strongly recommend that the reader either delete data sets created during the code development and testing process, or work in a production partition of the system that does not contain data sets created during the development and testing process. The reason this step is strongly recommended is that it is remarkably easy to accidentally input a development data set (that may contain incorrect information) in your production run. As long as your code is pointing to an existing data set, SAS will input and use the data set that the code is pointing to regardless of whether or not you intend for it to do so.

Project Disposition

Now that we have completed all of our programming, it is time to end our project. The disposition phase of the project is just as important as any other part of our work. As we discussed in Chapter 4, project disposition involves creating a "package" that will stand the test of time, documenting our work for future reference. In addition, because our project uses personally identifiable Medicare data, project disposition also includes the destruction of project data as specified in our Data Use Agreement (DUA). Should we perform this task improperly, we may not be able to turn to our work for future reference (which would be a real shame, considering all of the hard work we put into it!). Even more importantly, the consequence of not adhering to our DUA is something we do not dare to contemplate!

Archiving Materials for Future Use

Believe me, I have been here plenty of times. At this point in the project, you are probably very tired of the subject matter, as you have been toiling over a specific set of research questions and writing code. Nothing sounds better than to shut down your computer and go take a nap! I should warn you, however, that if you walk away now, you would be sorry. Picture this: Your fearless author, about one year after completing my first research programming project, is searching my hard drive in frustration for an algorithm I know is just perfect for my current project. Unfortunately, that algorithm is nowhere to be found, lost to time and buried in a jungle of files on my hard drive! You can learn from my mistakes by reading the remainder of this chapter and thinking about the best way to archive your work so it becomes a future resource, and not a source of frustration.

Our goal boils down to something quite simple; document and save the work we just completed in a way that enables us to utilize it for future work. In addition, CMS owns the code, logs, lists, technical documentation, and data created for any project funded by CMS, so archiving may involve additional work to transmit our archived materials to CMS. The following steps go a long way to completing our archiving goal:

- All programs should contain complete inline documentation, and should be clearly named so they sort in the order in which they run. For example, our code could be separated into distinct programs named 05_ETL.sas (the work performed to load our data performed in Chapter 5), 06_Enrollment.sas (the enrollment work performed in Chapter 6), 07_Utilization.sas (the utilization work performed in Chapter 7), and so on.
- All programs should be executed in batch mode so the logs and lists have the same name as their parent program. In addition, it is helpful to undertake a "production" execution of the code that entails running each program in sequence without breaks so that the date and time stamps on the files are sequential.
- All programs should be accompanied by technical documentation. Depending on your needs, technical documentation can be as simple as maintaining a listing of each program and its purpose, but as complex as an explanation of the purpose of every step in each program and including flow diagrams. Much of this book could be considered very comprehensive technical documentation for our code.
- All output should be labeled clearly and saved with meaningful filenames.
- All programs, logs, lists, technical documentation, and relevant output should be copied onto a CD-ROM or a DVD, being certain to follow the guidelines of our DUA. Including a "readme" file is helpful for documenting all of the files included on the storage media. It is important to keep in mind that the mere act of burning our files onto storage media does not guarantee that they will be preserved for time immemorial. Discs get scratched and hard drives break, so making a backup is highly recommended when contemplating long term storage of your work.

Data Destruction

Most DUAs require the destruction of all Personally Identifiable Information (PII) data and some additional materials derived from PII data, even if those materials do not contain PII data themselves. Therefore, the first step in destroying data is to become clear on exactly what data must be destroyed. At the least, we must destroy all files containing PII. For our example research programming project, we will destroy all SAS data sets. Next, we must choose a method to destroy the data. The reader should check the details of his or her DUA. Common methods of destruction include shredding CDs or DVDs, erasure via degaussing of magnetic storage media such as hard drives, and deleting data stored on hard drives, followed by overwriting the deleted files many times to ensure they cannot be recovered.

Chapter Summary

In this chapter, we closed out our project by performing the following tasks:

- We used the files created in Chapters 5 through 10 to present the results of our study.
- We discussed the process for undertaking a final batch execution of all of our programs.

- We discussed the execution of the disposition of our project by documenting and archiving project materials for future use and destroying our data per the project's Data Use Agreement.

Exercises

1. What additional research questions might you answer with the data created in Chapter 5 through Chapter 9? Can you select and present calculations in Chapter 7, Chapter 8, or Chapter 9 that were not chosen as an example for our work in this chapter (e.g., rates of chronic conditions by age or sex)?
2. This text studied a pilot program that was operational during calendar year 2010. Updating a study annually is a fairly common exercise in health services research. Therefore, perform a thought experiment based on imagining what we would need to change in this text if the pilot program was operated during calendar year 2015.

Index

A

B

D

E

J

L

M

S

T

U

V

W

Numerics

Lightning Source UK Ltd.
Milton Keynes UK
UKOW07f1607120816

280459UK00002B/112/P

9 781612 903224